5. Ore Team, Cunningham, Groff Coll.
6. Shenandoah Mine, Andreatta Coll.
7. Hercules Mine, Zanoni Coll.
8. Horse Winze, Zanoni Coll.
9. Frank and Alice Richardson, John Richardson Coll.

Mining the Hard Rock in the Silverton San Juans

A Sense of Place

A Sense of Time

Mining the Hard Rock in the Silverton San Juans

A Sense of Place

A Sense of Time

by

John Marshall

with

Zeke Zanoni

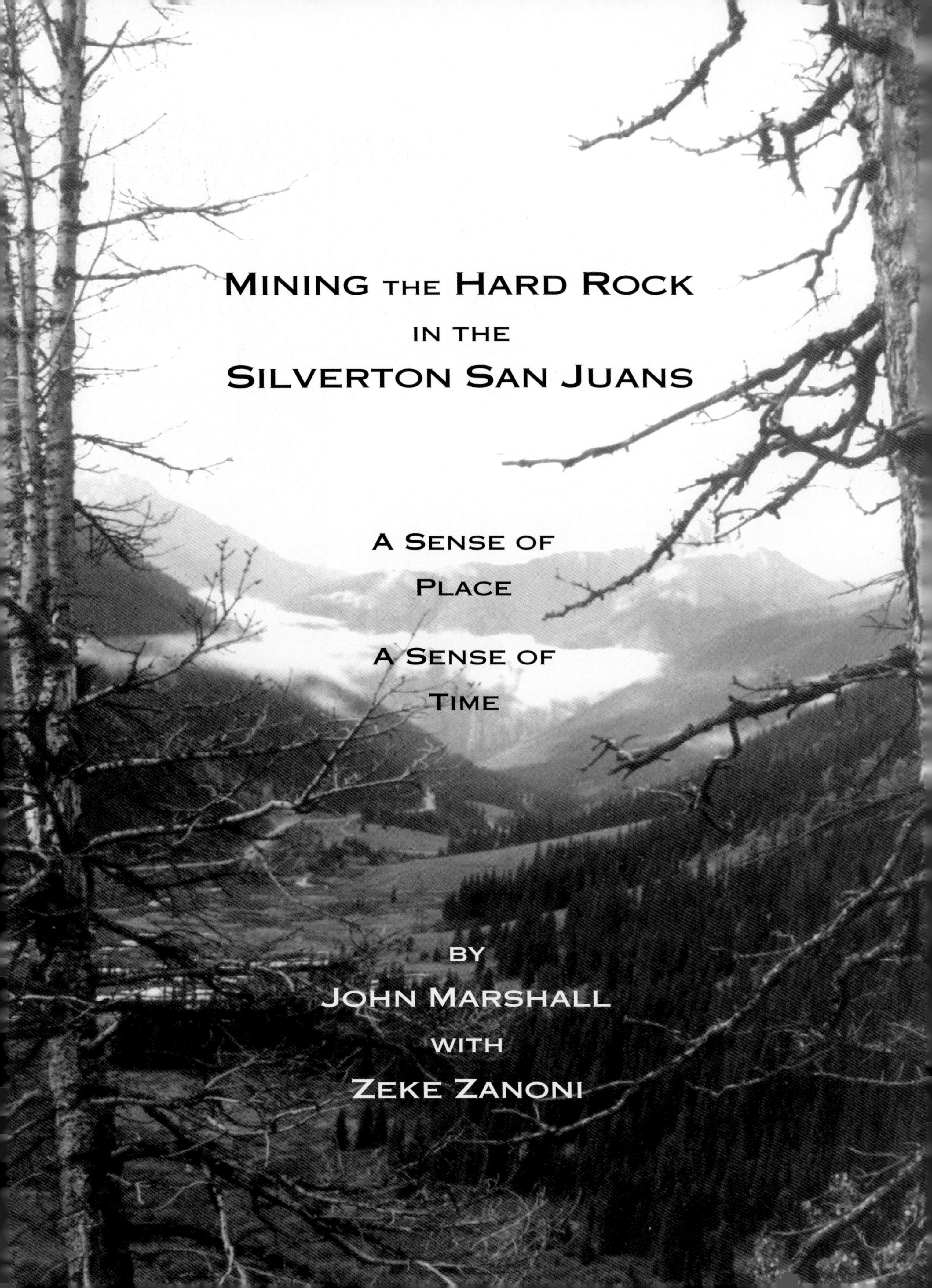

Other books by the Simpler Way Book Company:

A Simpler Way, Silverton Colorado
by John Marshall with Dan Mesich

Snowflakes and Quartz
Stories of Early Days in the San Juan Mountains
by Louis Wyman

Living (and dying) In Avalanche Country
by John Marshall and Jerry Roberts

MINING THE HARD ROCK IN THE SILVERTON SAN JUANS

Third Printing 2005

Packrat Publishing
P.O. Box 355
Silverton, Colorado 81433

BOOK DESIGN
SUSAN A. MATTHEWS
(pictured with daughter, Robin Erin)

Library of Congress Cataloging-in-Publication Data
Marshall, John.
Mining the hard rock in the Silverton San Juans; a sense of place a sense of time / by John Marshall with Zeke Zanoni.—1st ed.
p. cm.
Includes bibliographical references and index.
1. Mining History 2. Silverton, Colorado. 3. Colorado History 4. Oral Pictorial Americana I. Title
1996
Library of Congress Catalog Card Number: 96-94414
ISBN 0-9632028-2-0

Design by Susan Matthews, Serai Communication Arts, Durango, Colorado
Photograhpic laboratory: Mark Schuman, the dark room, Glenwood Springs, Colorado
Printed in the United States of America
A special thank you to Grace and Harvey Besaw for their permission to reprint excerpts from *Snowflakes and Quartz* by Louis Wyman, copyright © 1977, 1993. The stories are intended as a tribute to the native Silvertonian, now deceased.
Cover photo: Rainbow–Pot of Gold–Arrastra Gulch, Zeke Zanoni. Back cover: Collapsed Portal–Bausman Mine, Zeke Zanoni.

CONTENTS

Preface

When I started writing this book I didn't have any intention of writing a book on mining. I'd done some mining, I'd been around it a lot, and I'd seen it go through its cyclic ups and downs. No, I was going to write a book similar to my first one—about the people and town where I live, with an emphasis on photographs and short stories. There are a disproportionately high number of people with interesting and varied lives in my town willing to share with others. That's what the book was to be about. Then I started running into people who had worked the mines in the area. After all, Silverton is historically a mining town. Then many of these people started coming up with pictures of where they had worked—the buildings, the mines—of things that had disappeared, I thought, a long time ago. There was a small core of people who could still remember what had gone on and who had been a part of it. They were interested and even excited that someone cared. They became a part of this book. And these people won't be around forever. Now, in these modern days where gold's primary value is for jewelry, economics will force them to move. Age will catch them, and change, yes even in the small, isolated town of Silverton, will overwhelm them. So when Zeke Zanoni was willing to throw in with me I knew that this was going to be a book on mining. A book about change. *Mining the Hard Rock* was a learning experience for me. I hope it is for you too.

Dolores LaChapelle Photo

The Author,

John Marshall

Oh yes, I've worked in the mines—been paid to go underground. Never did break any rock, but I surveyed and sampled and worked with an exploratory drilling crew—in the dead of winter with avalanches all around. Boy! The older I get the better I used to be. And I've worked in the mill, in the crushing plant, and on the ball mill. I know what it's like to be dirty, dead tired, and cold and wet. So now I'm trying to write books. Is it any better? Maybe don't ask...

ZEKE ZANONI

John Marshall Photo

A special note of thanks to Zeke Zanoni. Without his help this work on mining would not have happened. He delivered patience, a wonderful collection of mining pictures, sketches, hand-drawn maps, and a willingness to share his knowledge with someone who isn't a miner, making this book far richer than I had foreseen. His grandfather Ernest was a Silverton miner as was his father Leach. Zeke, like his brother Tom, is also a miner, I guess you could say mining is in his blood. Zeke was kind enough to make himself available for dumb questions, write sections when my knowledge failed, make drawings with his sound artistic ability and mining wisdom—to the point that sometimes I thought he was writing the book. As the work was being finished, he was glad to read it through, meticulously, and offer corrections and criticism. Without his help and friendship this book would have fallen short of the mark. I'll take the blame for the mistakes but when it rings true, know that somewhere, somehow, Zeke Zanoni had a hand in it.

Other old-time miners such as Joe Todeschi, Vince Tookey, Rich Perino, Tom Savich, Bill Rhoades and a miner's wife, Wilma Bingel, and the younger folk like Mike Andreatta, Pat Donnelly, Bill Jones, Scott Fetchenhier, John Wright, Jim Melcher, Lois MacKenzie, Christine Bass, Beverly Rich, as well as numerous others like Gerald Swanson, Herman Dalla, Allan Bird, Ruth Gregory, Andy Hanahan, and Ernie Kuhlman, and others too many to mention, were invaluable with their pictures, knowledge and stories.

"एउटा ठाँउको भावना"

A sense of place

"एउटा समयको अनुभव"

A sense of time

Molas Lake, 1994. ***John Marshall Photo.***

It used to be a pretty quiet place with just a handful of summer visitors. All Indians.

The side of Kendall Mountain. *John Marshall Photo.*

The moon would rise
over a land of snow and ice,
flowers, birds and animals…

The prospector. "In the mining game, gold is just where you happen to find it. And you never know what's under your feet." Tom Walsh, 1896. ***Photo Courtesy San Juan County Historical Society.***

But things change,
at first so subtly…
Then so suddenly!

"But events such as these are just the vehicles change likes to ride around in. Evolution drives a bulldozer disguised as a stationary bike. With history, it's the other way around." Tom Robbins *Half Asleep in Frog Pajamas*

—a queen city of the western mining towns in the early 1900's, high in the mountains, constantly growing, constantly changing, yet subject to the vagaries and whims of economics and politics and disasters, natural or otherwise…

Tom Savich Collection

BREWERIES, SMELTERS, SAWMILLS, FOUR TRAINS A DAY

—running from Durango to Silverton, running from Silverton over the top of Red Mountain on to Ironton, running up Cement Creek to the little town of Gladstone, up Cunningham, out to Howardsville, Eureka, Animas Forks… running, running, running!

Two views of the Silverton Brewery—the town's answer to Fischer's Rocky Mountain Brewery at Howardsville. The brewery in Howardsville had been the first one on the western slope and Charles Fischer eventually came to Silverton to run this brewery. Having burned once, it was rebuilt with stone. Beer was sold in fifths or by the barrel.

(below) The brewery was in back, the offices in front. In winter they used the pond for ice. On summer Sundays in 1909 the townspeople would flock to the pond for picnics, maybe some fishing and free beer.

Rock from the building at the left was used in construction of the shrine on Anvil Mountain (see page 151). Eventually, the rest of the structures were used for highway fill at the first pullout on the road to Durango, just at the edge of the city limits.

The smelter billowing smoke in the background was the Martha Rose-Walsh Smelter. Smelters were short-lived here primarily because they don't work well at such high altitude. Nearby, a sawmill produced shingles and board feet of lumber numbering in the hundreds of thousands.

Denver Public Library Western History Collection

Smelters sat at both ends of town, this is the Kendrick Gelder Smelter above Cement Creek. Note the tracks going to Gladstone at the bottom right of the picture. *Colorado State Historical Society, Wm. H. Jackson Photo*

The train stopped on top of Red Mountain before continuing on toward Ironton. Elizabeth (the baby), Jeanne and Hazel Root, along with Mr. Norton, are disembarking from the train. The station would be to the left, but the steps look in disrepair. The year was probably around 1910, and indeed, this was the Rainbow Route. *Cynthia Fransisco Collection*

During these prosperous times construction flourished. There were plenty of wooden structures, and many of these stand today. The first brick buildings were built by bricklayer Patrick Stanley in 1877. One was on Reese Street, one by Cement Creek just off Greene Street. The Posey and Wingate Building at 1269 Greene Street is probably the oldest commercial brick building in Western Colorado, built in 1880.

Silverton after the peak year of 1907. View facing northwest towards Anvil Mountain. *Gerald Swanson Collection*

Quality Hill. One could view Blair Street, the notorious red-light district, from here. *E. Adams Photo, Judy Graham Collection*

Brick and stone seemed to be a sign of stability. These buildings listed were some of the first made of brick or stone. The town might fall down, but it wouldn't burn down.

1880 The Pickle Barrel Restaurant
1883 The Grand Imperial Hotel
1893 San Juan Cafe
1896 French Bakery
1901 The Benson block (commercial)
1901 American Legion
1902 Wyman Hotel
1902 County Jail
1902 Alma House Hotel
1904 The Avon (hotel)
1905 Saint Patrick's Catholic Church
1906 The Library
1906 Exchange Livery
1907 Miner's Union Hospital
1907 San Juan County Courthouse
1908 Town Hall
1911 Public School (grades K through 12)

This fine example of beautiful brickwork was constructed around 1905 and still stands by the library, on the corner of 11th and Reese.
Terry Kerwin Collection

Louis Wyman set up a summer camp on the Rio Grande on the Creede side of Stony Pass with family and friends—a normal summer procedure for the Wymans. Dr. Mechling, a long-time family friend, drove this car from Denver to Creede and over the Continental Divide at Stony Pass (elevation 12,588 feet, on a road that was never better than a pack trail) to the camp for an historical rendezvous at the end of August. Mechling's son Tug had spent the summer with the Wyman family. Here, the two men and their sons are in sight of the top, chains on and fingers crossed. *Grace Besaw Collection*

The First Car

Louis Wyman, as a 10 year-old boy, went on the trip of a lifetime with his dad, Louis Wyman, Senior. Louis later recorded the journey for us in his story, "The First," excerpted here from the book Snowflakes and Quartz.

"Dad's daydreams were not the idle musings or the fantasies of a mystic. They had a way of erupting into reality. He advocated and built good roads. During the summer of 1910 he was up to his sweatband in a project that took in half the State of Colorado. He envisioned an automobile highway, unheard of at that time, running southwest from Denver through the scenic grandeur of the High Colorado Rockies. The road would traverse a thousand miles or more of the state's finest recreational land, finally circling back to Denver. Years later, when all the county roads had been connected, it became known as 'Colorado's Thousand Mile Circle Route.' Of course, Silverton and the San Juan were bright stars along the way."

Louis Wyman and his lifelong friend, Dr. D.L. Mechling undertook the outrageous plan to drive a car from Denver to Creede, up over Stony pass and down Cunningham Gulch into Silverton.

"The car was of the Croxton-Keeton type, 30 horsepower. Four cylinders – 4° inch bore, 4° inch stoke – a

Louis Wyman, Senior, was an adventurous opportunist—not afraid to initiate change. He was a packer who sold his stock and equipment to the British for use in the Boer War in South Africa before 1900 and then built the Wyman Hotel in 1902. He knew that Silverton needed a car. *Grace Besaw Collection*

car patterned almost exactly after the French Renault." Even so, it took a skinner named Gil and his team of horses to get that car up over the last steep pitch of Stony Pass. Numerous crossings of the Rio Grande River on the way up caused a lot of problems, constantly having to remove and dry a wet distributor cap. Then there were flat tires and finally they had to chain up through summertime mud. But, at last, the summit was achieved and the rest of trip would be all downhill—but downhill had its own challenges.

"Stony Creek Gulch lives up to its name in every respect. From the summit down to where it joins Cunningham Canyon is a steep and rugged defile, dropping 2,500 feet in less than four miles. On the west side of the Continental Divide the wagon trail had been improved slightly because of mining activity. It was never, however, intended for automobiles. We worked our way down slowly, stopping now and then to roll boulders out of the way and cool the brakes. Once they had to be tightened. But by four o'clock Uncle Mech was jockeying the car around the last hairpin curves of Stony Gulch, and the car rolled out onto the county road in Cunningham Canyon.

"Silverton awaited us. Word had been telephoned in from a mine that the car was down off the mountain and headed for town. It seemed to me that everybody in town had turned out to celebrate the ushering in of a new era and to welcome the first automobile to the San Juan and Silverton.

"Dynamite exploded on a hillside above town. The band was assembled in front of City Hall, doing their best to be heard above the wild cheering, as Uncle Mech braked to a stop with a flourish and a long blast on the horn. There were people in that crowd who had never seen an automobile except in published pictures." And maybe some people who had never seen one anywhere. But there it was—the automobile. It had reached another place, probably the highest spot any car had ever reached, in America. It was an exciting time and in the years that followed only a few people wished for quieter times as they had once known them.

From *Snowflakes and Quartz, Stories of Early Days in the San Juan Mountains* by Louis Wyman.

The town of Silverton turned out for the first car ever in 1910. *Grace Besaw Collection*

And then it was on to Ouray. In this photo they are passing the more conventional horse and wagon at the outskirts of Chattanooga, just below the top of Red Mountain Pass. *Grace Besaw Collection*

The War Steals the Young Men Away

From United States Supreme Court Justice William O. Douglas' fine book Of Men and Mountains *comes this excerpt. I'm sure his words spoke for all young boys across America when it came time for them to go to war. These very thoughts were on every person's mind right there in Silverton as they stood on the platform by the depot about to depart on a journey into the unknown.*

It was late summer in 1914. I was fifteen and entering my junior year in high school that fall. School would soon open. I saw a chance to get away for a week and took it.

Herman Dalla, born in 1912, is warmed by the stove in the county shop as he reflects on the rough periods of Silverton's history having lived through prohibition, war and the flu.
John Marshall Photo

As I was rolling up my horseshoe pack, I had an idea. I would be gone a week or more. I planned to enter the Cascades through the Naches and the Teton and go back into the lake country beyond Cowlitz Pass. I would most likely run into a sheep herder in that region. The papers carried big news—news of war in Europe. So I decided to take recent issues of the Yakima Daily Republic with me. I put several in with my blankets and also a recent issue of the weekly magazine, *The Outlook.*

A few days later I was skirting a meadow on the east side of Cowlitz Pass when I saw fresh sheep droppings. I left the trail and went south across the meadow. The sheepherder who greeted me was a middle-aged man with a full brown beard. His eyes were blue and kindly. His face was bronzed. He looked a little as I imagined Walt Whitman must have looked. This man was tall, long legged, long armed; a wiry, rangy man who appeared to be equal to any challenge of the mountains. His voice was resonant, with powerful carrying qualities even in ordinary conversation. It was a voice that came back to me four years later as I drilled and marched in the uniform of the United States Army.

"Hi ya." He invited me to his camp and he invited me to dine with him. "I haven't seen a paper in four months," he said. "So you will have to bring me up to date." I handed him the Yakima papers and the copy of *The Outlook.* He thanked me, and after a pause said, "Will you do me a favor? Read me the paper while it is still light."

So while he cooked supper, I read the most recent paper. Most of it was news of the war—the Kaiser, the Huns, the English Channel, Flanders, the Tricolor, the Marseillaise, the Rhine.

It was deep dusk when I finished the first paper. Supper was about ready. When I started to eat, he left with his dogs to tend the sheep. He was back in an hour or less. As he was filling his own plate, he said, "Son, do you think you can see to read me some more?" So I lay on my belly by the campfire and read on and on as he ate.

It was a still, clear night. There was a touch of fall in the air. I stayed close to the fire. He cross-examined me.

Key: 1. Father Berry 2. Bill Eniss 3. Joe Dresback 4. Paul Hoffman 5. Lew Hawes 6. Martin Slabonick 7. Jimmy Billing 8. Fat Fleming 9. Pete Dresback 10. Art Sullivan 11. John Pentalone. *Gerald Swanson Collection*

He not only wanted to know about the war; he wanted to know about baseball, the price of hogs, sheep, cherries, and hay, the news of the valley, of Woodrow Wilson, Congress, Teddy Roosevelt and Pershing. I could not answer all the questions he asked, though I tried to give him a synopsis of events during the summer of 1914.

After he had pumped me dry, there was silence. There was not even a murmur in the tops of the firs that guarded the camp. There was the crackling of the fire and the faint sound of the snoring of one of the dogs. All else was quiet. The stars hung so low that they almost touched the firs.

The war in Europe? It was as remote as the typhoon that swept bare an island in the South Pacific whose name I could not even pronounce. War in Europe? That should not concern anyone here. Hasn't Europe always had wars? Even a Hundred Years War? The war was remote, as foreign as a flood in China or a revolution in Persia.

That is why, I think, the evening in the meadow below Cowlitz Pass remains so vivid in my mind. For as I sat in silence thinking of the war as something wholly removed and apart from our world, the sheepherder spoke, "Well you boys may have to finish this."

How right he was. The boys by the Silverton Depot would learn this. And there wasn't much available in the way of good choices. One year later, those who didn't go to war would face the deadly flu at home.

Silverton Struggles Through Prohibition

Prohibition started in Colorado on January 1, 1916 by the good people in Denver who, even then, thought they knew what was best for the entire state. By 1917, an agent from Durango would visit Silverton fairly regularly.

Those agents busied themselves with more than the obvious and expected visits to the many saloons. (Were there bribes and loose lips?) There were a lot of talented people living in town from various countries around the world, and the making of alcoholic beverages was often an old family tradition.

Take the Italians. Domenica Todeschi Dalla and her husband John had eleven children. To help pay the bills, they ran a boarding house, part of which was a saloon. Producing wine was a natural part of their livelihood. The same was true for many others in town as well.

A thousand gallons of wine could be made at a time. Herman Dalla once remembered three boxcars full of grapes coming in on one train. So when prohibition came, there wasn't going to be any closing of saloons—whiskey was simply added to the list of homemade liquors. Herman's brother spent many an hour tending stills and mixing elixirs in the unused corners of dark, quiet basements spread throughout the town.

Business was good. There was a constant demand for alcohol as well as other entertainment. Blair Street was two blocks long and ran from 11th to 13th Streets. It cost $2 to $3 dollars for a fling. Every saloon paid its monthly tribute of $25 dollars, and every prostitute paid $14.85 to the city for disturbing the peace. The big saloons were known to pay in the neighborhood of $125 dollars to 'sources' every three months; they might get raided, but nobody would ever find anything. About this time there were two banks, two hardware stores, three movie theaters, twenty-six saloons, five grocery stores, 2,000 people and a train you could get on in Animas Forks and ride to Denver. But already the busy times were winding down. Silverton's heyday was on the downhill slide. Prohibition would allow fortunes to be made in the sales of beer, wine and whiskey, but it would also mark the start of events that would put Silverton into slower, quieter times.

Amendment 18, calling for national prohibition, was ratified by the federal government on January 16, 1920. Besides Colorado, twenty-four other states already had their own prohibition in place. For fourteen years prohibition reigned, until December 5, 1933, when Amendment 21 repealed the law that had gotten little respect.

Back home in Silverton in 1928, revenuers once confiscated 2,600 gallons of mash. They took it down to the end of town and dumped it in a pasture. The cows thought that was fine and dined happily. However, for the next three days all the milk had to be thrown away for the cows had gotten drunk, and the milk wasn't any good.

Annie Smith, 90 years old in 1995, suffered in her own way during prohibition: "I quit school when I was twelve years old because of those liquor laws. I'd gone to Durango on the train. In Durango, with my father, we were sitting in a room, when a revenuer [more commonly called a *revenooer*] burst into the room quite unannounced. There was a bottle sitting on the table. I was just a kid, but I grabbed the bottle and threw it out the window. We heard it break on the sidewalk below. The agent took me and my father to jail. After three days, one of the conductors on the train heard about it. He came and got us out of jail. I rode the train back to Silverton. I never went back to school, and I didn't even come out of the house for almost the rest of the summer." ❋

As much as anyone Annie Smith and Corky Scheer represented the times of which we write having witnessed the hardships and devastation of war, prohibition and influenza. By 1995 they both had passed on. *Photo courtesy of Louis Smith and the Joe Girodo Family*

Ernest Allen · Joe Anderson · Kristine Anderson · Nels Andren · Joe Angella · Frank Babich · Jim Barchetta · Joe Barnardi · George Basarich · Alta Bausman · John Bertolodo · M. Bettes · Fred Blackler · Angelina Bonovida ·

Pete Bragnara · Wilma Buck · George Burgoyne · Genevieve Clifford · James E. Cole · Frank P. Corley · Herbert Crane · John Cura · John Curadi · Pete Curadini ·

Albino Dalla · Anthony Dalla · Domenica Dalla · Matt Dalla · Fendor Drakulick · Edward Eales · Pearl Eastman · Gegee Eccher · Albin Erickson · Carlo Erspan · John Fedel · Pete Ferrari · Mrs. Addalay Gallegos · Jim Garolene ·

Pete Garoni · Angelo Giacomezzi · Dominic Giovannini · Giuseppi Giovannini · Linda Giovannini · Nicklo Glumac · J.E. Gorden · Berne Haas · Carl Hansen ·

Then Came the Flu Very Quickly and Very Deadly

HARRY. HE NEVER SEEMED TO CHANGE MUCH. Kind of like an old fence—if a picket or two falls off nobody notices. I can't recall ever seeing him dressed in anything other than a blue-serge cap, dark shirt, paint-spattered pants and vest. He worked out of a local hardware store, as a plumber, tinner, and man of all trades.

It was only when some of our good townspeople were in trouble, or had a dirty job, that they remembered Harry. He replaced their broken windows, fixed their leaky water pipes, put the new stove grates in for them and took the stoppage out of the bathroom seat after Junior had plugged it up. Whatever had to be done, he could manage.

A few beers or a pint of wine on Saturday evening was about the extent of his social life. None of it ever showed in his behavior. The smile and kindly eyes squinting from under his cap bill were always the same. Along with his Cockney accent, shuffling walk and hunched shoulders, he seemed as durable as an old pine-knot.

In the fall of 1918 the flu struck our town. For a while people fought the plague with spirit and courage. But their defenses crumbled as the death toll mounted. A time came when the dead could no longer be cared for in the usual manner. No help was available from other communities. They, like Silverton, were prostrate. Prepared food for those who could no longer care for themselves arrived by train daily. The Town Hall became a hospital, or, more aptly, a place to die. It seemed the black crepe of death was hung on every door in town.

Harry forsook his little shop behind the hardware store. Through those dark days he labored around the clock in the morgue. After the supply of coffins was exhausted, he made rough boxes for the dead or wrapped them in blankets as best he could before they were taken to the cemetery.

If he felt fear of the plague, he mastered it and lived with the dead. As long as there was a body to be cared for, he stayed at his self-imposed duty. His only rest was a short nap in a chair until called again to help a stricken family care for their dead.

Harry never wore a flu-mask, nor took any other precautions to protect himself against the infection. He was far too busy helping desperate people through their days of despair. If he wasn't needed at the morgue, he went to homes where families were down and helpless. Many a one walked the streets again because of him.

When it was all over, and the town had shaken off its terror after that winter of flu, Harry was back at work. He went shuffling along with a window pane, a rubber sewer-pump or a Stillson wrench tucked under his arm. Things were getting back to the way they should be, and no one doubted that Harry would be there to do the patching and fixing.

How do you measure the worth of a man? Is it by the castle he builds? By the battles he fights? Or perhaps by the wealth he amasses? I don't know, but I think the rod would have to be long to measure a little guy who stood tall and said to his neighbors, "Come, let me lend you a hand."

Harry Rogers was born in Silverton in 1882. His father and mother had come to town some time before 1876 from Cornwall, England. Harry remained a bachelor and resided with his mother on the ground floor of the Masonic Temple on Reese Street. Harry died in 1953 in Montrose, Colorado.

This story excerpted from *Snowflakes and Quartz, Stories of Early Days in the San Juan Mountains* by Louis Wyman.

Sam Hennon · Art Hill · George Hitti · Carl Holmberg · Pearl Bessie Howe · Perry B. Howe · Richard Hughes · Ernest Jackson · Carl Johnson · Ed Johnson · Gus Johnson · Rasmus Johnson · Davy Jones · Louis Kochever · Bosko Kovacevich · John Krappo · H.S. Lee · Frankie Leonard · Louis Leonardi · Tony Lilsole · Mary Lira · Ray Lopez · Margaret Lorenzon · Pete Macconi · Joe Marches · Al Marshall · Diego Martinez · Mrs. Nick Martinez · Aresto Matavi · John Matress · Angelo Mattivi · Dan McCloud · Floyd R. McCoy · George M. McCoy · Richard Earl McLeod · Margaret McNamara · May McNamara ·

THERE ARE STILL A FEW PEOPLE AROUND WHO REMEMBER PROHIBITION, WORLD WAR I AND THE FLU. Throughout the following text excerpts from the *Silverton Standard and the Miner* and personal memories recount these gruesome times.

October 12, 1918–In most cases a person taken sick with influenza feels sick rather than suddenly. He feels weak, has pains in eyes, ears, head, or back and may be sore all over. Many patients feel dizzy, some vomit. Many patients complain of feeling chilly, with this comes a fever of 100° to 104°.

Everyone is ordered to notify a doctor on the first signs of any symptoms of sickness and to use every effort to prevent any spreading of the disease.

October 19, 1918–Silverton celebrates false rumor of Germany's surrender. Parades and bonfires go through the night.

Mickey Logan, now living in Durango, still operates his electrical business at 91. Mickey was born in a house in Silverton, October 19, 1904 in a raging snowstorm. "I didn't like the snow that day, and I haven't liked it since!":

"My dad, Bill Logan, ran a freighting business in town. He got wood from down by the river and hauled it up to the intersection by the Pickle Barrel. A big bonfire was built, and there was music and dancing all night long. Bottles were shared and gaiety reigned. But by daylight the flu was killing the celebrants."

October 26, 1918–The Worst Week Ever Known in San Juan County

November 2, 1918–128 dead. In two weeks time Silverton has lost a number of her most prominent people, men and women in all walks of life have been taken from us in death. In no city, town, or village has the epidemic of Spanish influenza proved more fatal.

The Miner's Union, the Legion, used to be a mortuary. R.F. "Bud" McLeod had bought the business from Horace Prosser in 1912. The bodies were piling up. Bud contracted the flu and on October 28, he died within a few hours of his sister. Big, burly, Slavic miners would sit down at lunch, strong and healthy. The next day their chairs would be empty.

Mickey Logan and Johnny Mac (McNamara) in Mickey's electric shop in 1994.

John Marshall Photo

Thomas McNamara · Charles Melton · John Mesmis · Jim Micholetti · Louis Micholetti · Bessie Miller · Edna Jane Miller · Evariste Montoya · Celeste Motto ·

Killed by an unseen bug. Soon there were more than 100 bodies in the mortuary. Something had to be done. Two trenches were gouged along the lower road of the cemetery in the cold winter-hardened ground. The placement of the bodies was recorded. Later, they would be disinterred and moved to the cemetery proper, where services were rendered.

Throughout the mining district activity has been suspended during the epidemic.

Chester Black came up from Durango to be the mortician. Bill Maguire helped out and by 1927 had became the County Coroner. He remained an embalmer until his retirement 45 years later in 1972. His son Bill Maguire, Jr. resides here today.

November 16, 1918–The signing of the real Armistice resulted in no celebrating in Silverton.

November 23, 1918–The pool rooms reopen, as do the picture houses. The school follows suit, and the library…

It was a time of much personal tragedy. "The front of our boarding house was a saloon. Hell, I was only six years old but I can remember the wagon coming from the mortuary and them loading my mother's body into the back. I was watching from an upstairs window. Two of my brothers died. That left nine of us kids with no mom and no dad. Mary was the oldest of us and she kind of took over. We had records of who owed money to mom and the boarding house, but everybody said they'd paid up. It was rough. When I reached sixteen I was on my own."

"What else do you remember, Herman?"

"I remember big guys—over six feet. Husky. Strong! I'd remember them sitting down to supper and in the morning they'd be dead. Oh sure. People wore rags over their faces. I don't know if it did any good or not."*—Herman Dalla*

"None of the Anesis got sick. They would take groceries to the houses of the ones who were ill. They would make soup and bring it around the town. No, they didn't wear masks."*—Louis Smith*

"I can remember Dad was hauling bodies to the cemetery in his wagon, four or five at a time. The doc was Doctor Burnett, I believe, and the druggist was Mr. Stewart. Well, they'd give me the prescriptions and I'd go deliver them. I wore a mask across my face. I'd go in their house and if they didn't have a fire going, I'd build them one. Then I'd haul some kindling and some wood for them. Some of them lived and some of them died. No, I never did get sick. My dad, neither. And he never did wear a mask. I guess the Logans were lucky this time."*—Mickey Logan*

"Bessie Rivers was a society girl from a mainline Philly family, so the rumor went, and she showed up in Silverton one day to become a madam. Now, in 1918, her younger sister came to visit, worrying Bessie that her classless occupation might soon be found out and told to the family back east. But as soon as her little sister showed up, she caught the flu, and that worried Bessie even more. Well, my grandma and her daughter (my mother) took over caring for this stranger from the east. With a little time and much luck and tender care, they were able to nurse the young girl back to health. Bessie got Sis on the train as soon as she was strong enough to travel and thus her occupation was kept a secret. In appreciation of my Mom's and Grandmother's care, Bessie one day slipped her a little black velvet bag. It was full of diamonds. They were passed on through the family and thirty years later some of them were made into this beautiful platinum diamond ring. Days of caring and sharing…"*—Gertrude May* ❋

A ring of thanks… **John Marshall Photo**

YET THROUGH THE TRIALS, MINING SURVIVED...

In November of 1918 the war with Germany was ending. Elections were held. In Silverton friends and neighbors all across town were sick and dying. The flu turned into spinal meningitis and pneumonia. Yet, Thanksgiving was coming too. Those who were alive had reason to be thankful. And there was mining. And it was mining that was on the front page. Through sickness and war. Mining was the beginning of Silverton. And though over the years it would struggle and sometimes even quit, mining would carry Silverton through the Great Depression of the thirties and on into the eighties. But would it last forever?

Bill Rhoades, forty-five years a miner, holds a plaque of appreciation and relaxes in his yard in 1994. He'd received official paychecks from at least fourteen mines.

John Marshall Photo

"Lord, how I loved the mining. There's nothing else I've ever done in my life I'd trade it for. To be called a miner, why it made me feel proud. For a miner could do it all. And he learned from the guys he worked with. They showed him the skills he needed. I'm proud I'm a miner."
—*Billy Rhoades, 1994*

These are shaft miners on top of Little Giant Mountain. Note the electric light in the background. Probably around 1905.

Chase Family Collection, William R. Jones

AND THAT'S THE WAY IT WAS. Rough men in tough places. Proud of what they could do. Proud of what they did.

This man is high on Little Giant Mountain, probably in the early 1920's. Arrastra Gulch drops away to the left. He's running a sinking drill that was used to make shafts. *Chase Family Collection, William R. Jones*

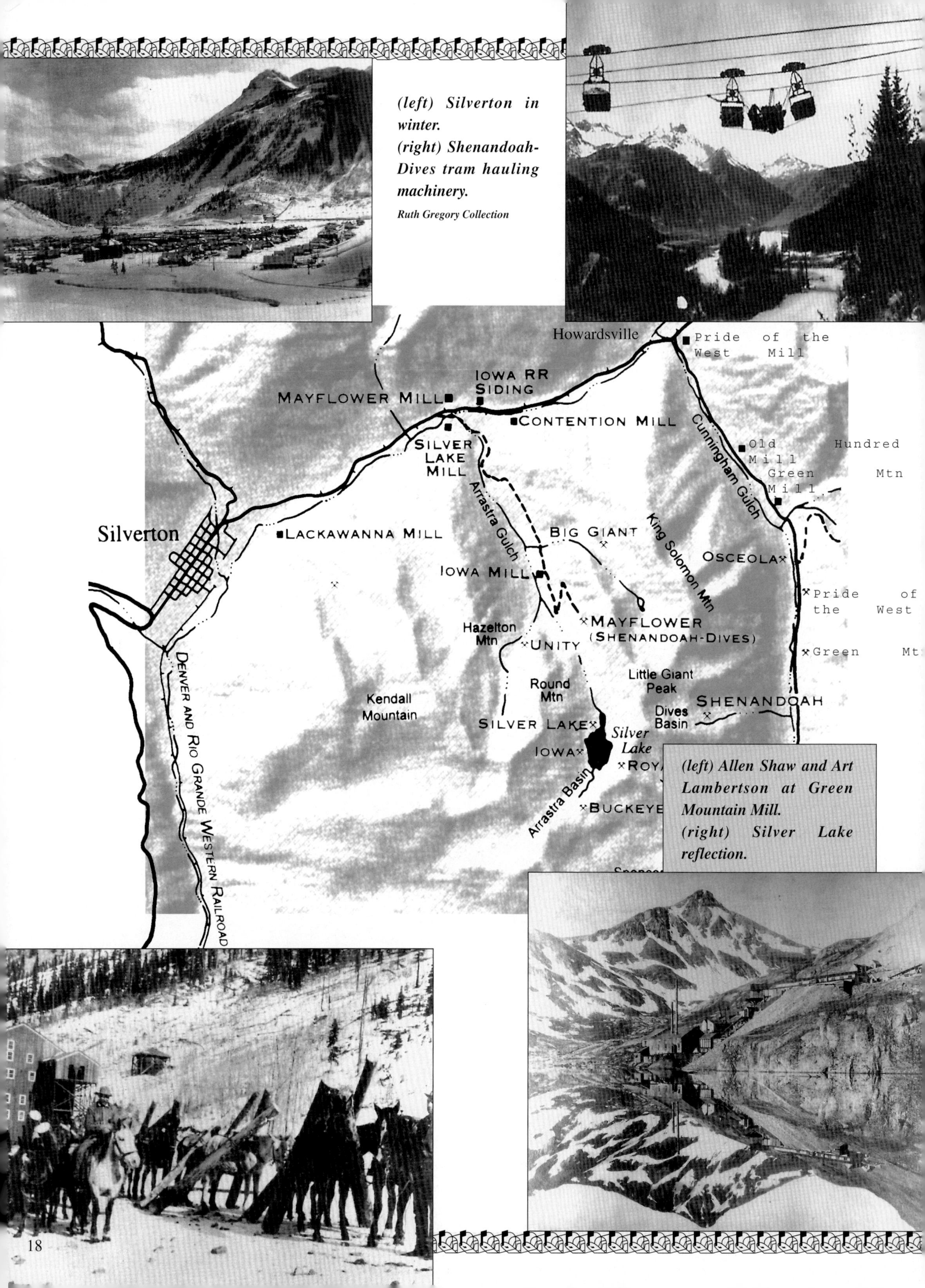

(left) Silverton in winter.
(right) Shenandoah-Dives tram hauling machinery.
Ruth Gregory Collection

(left) Allen Shaw and Art Lambertson at Green Mountain Mill.
(right) Silver Lake reflection.

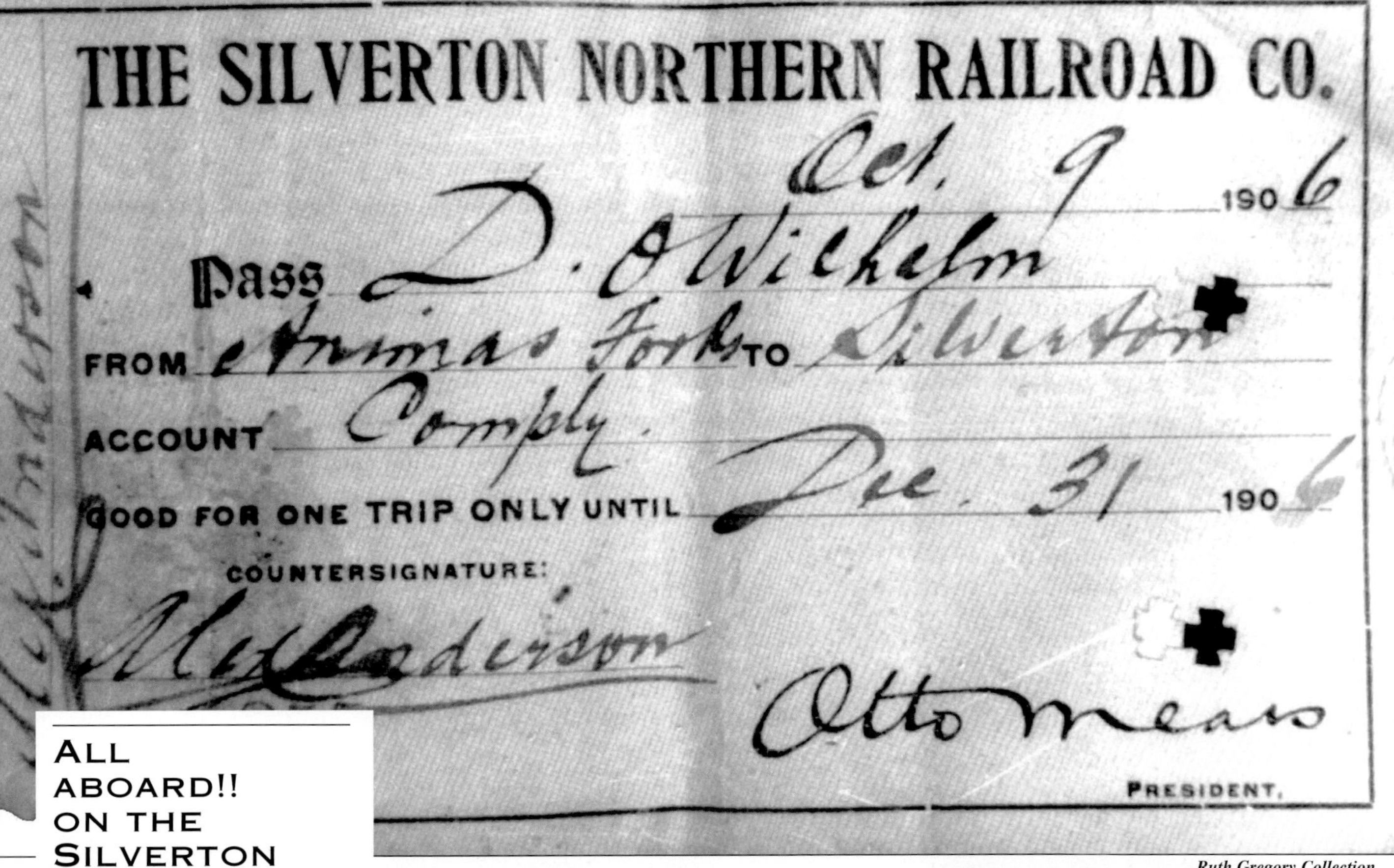
THE SILVERTON NORTHERN RAILROAD CO.

Oct. 9 1906

Pass D. O. Wilhelm

FROM Animas Forks TO Silverton

ACCOUNT Comply.

GOOD FOR ONE TRIP ONLY UNTIL Dec. 31 1906

COUNTERSIGNATURE:

Otto Mears

PRESIDENT.

Ruth Gregory Collection

All aboard!! On the Silverton Northern First stop Silver Lake Mill

Well, c'mon now, let's play our cards right, take our chances and head out of town on the train and see what's going on in this world of mining. After all, the train was the easiest way to get around in the early days. We'll have to wander in time a bit as well, but since it doesn't look like we're playing with a full deck anyway, that should be easy.

We're going to visit the mines. Without the mines and the miners there may never have been a town…

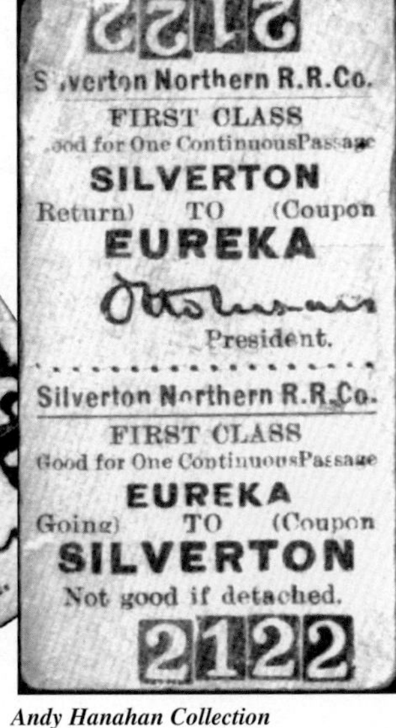

Andy Hanahan Collection

Adam Smith, Dotty's little boy on Blair Street— age 6 months, 1974.

Tommy Wipf Photo

THE MINES OF ARRASTRA GULCH

In the picture below, with Galena Mountain in the background, is the complex of the Silver Lake Mill. It is located near the bottom of Arrastra Gulch by the Animas River and the railroad. The Silverton Northern is shown coming out of the spur at the mill. The ore for the mill came down from Arrastra Basin and into the angle station of the aerial tram located next to the Iowa Mill. Historically, Arrastra Gulch is from Silver Lake Falls above the Shenandoah Mine down to the Animas River.

Arrastra Basin is the higher area near Silver Lake. The Silver Lake Mill was run by Lena and Ed Stoiber who built the Stoiber Mansion known as Waldheim. Lena was locally nicknamed Captain Jack, possibly for her habit of driving downtown in her wagon to round up her men out of the bars and cart them back to work at the Silver Lake Mill. She also drove the golden spike on the Silverton Northern Railroad in Eureka, June 1896.

Andy Hanahan Collection

The tracks are gone and only the bones of the Silver Lake Mill remain at sunset in July of 1994. *John Marshall Photo*

Let's jump off the train and head up into Arrastra Gulch to the Iowa Mill. There's a jeep trail here where the aerial trams once ran. It's a pretty spot in 1994, pretty darn quiet. But it wasn't always that way.

The busy sounds of the Silver Lake Mill are still loud in our ears as we start up the trail. An empty tram car clatters over our head on its way back to the mines of Silver Lake. The thick, black forest engulfs us. The sound of a train whistle filters through the branches. Shouts of some men cutting timber sound out. We continue upward. Then, we step off the trail to let a string of pack animals by, their bells tinkling, lumbering down from the mines above. What a busy place! Now, the grumbles of the Iowa Mill start to reach our ears as we climb higher.

On the left, Little Giant Peak, 13,416 feet. On the right, Round Mountain, 12,912 feet high. Silver Lake lies over the ridge between them. *John Marshall Photo*

But time has passed, the sounds have faded. The men are gone. Empty buildings stand against the weather. An unsecured door bangs mindlessly against the wooden side of a building once full of working machines run by men earnestly pursuing their dreams. Now, the countryside is quiet. Arrastra Creek, its gentle flow broken by mining debris, no longer tells its tales to passing miners. To be sure, an occasional lessee may come into the gulch and for a while a mine will run. But the boom is gone—the glory years are over.

It was into this once-busy valley that Charles A. Chase came in 1925. He came with knowledge and financial backing, searching for a place to begin a new mining enterprise.

The Iowa Mill, located half way up Arrastra Gulch, as it looked around the turn of the century, 1902. It processed ore from the Royal Tiger Mine and the Iowa Mine located on the shores of Silver Lake. This mill would also process the ores of the Shenandoah-Dives Mine for the first two years of that mine's operation, while the new Mayflower Mill was being built down in the Animas Valley. *John McNamara Collection*

The Iowa Mill, again around 1900. Behind it, to the left, is the Angle Station, so named because it could change the direction of the incoming trams. The ore from the Silver Lake Tram and the Unity Tram passed through here on the way to the Silver Lake Mill down below by the railroad tracks. In 1902 this mill was sold to the Guggenheims and the American Smelting and Refining Company. It ran sporadically until the early 1930's when the Shenandoah-Dives group became the dominant force in Arrastra. The mill was located where the avalanche-crushed Quonset hut sits today. Zeke Zanoni Collection

Arrastra Gulch

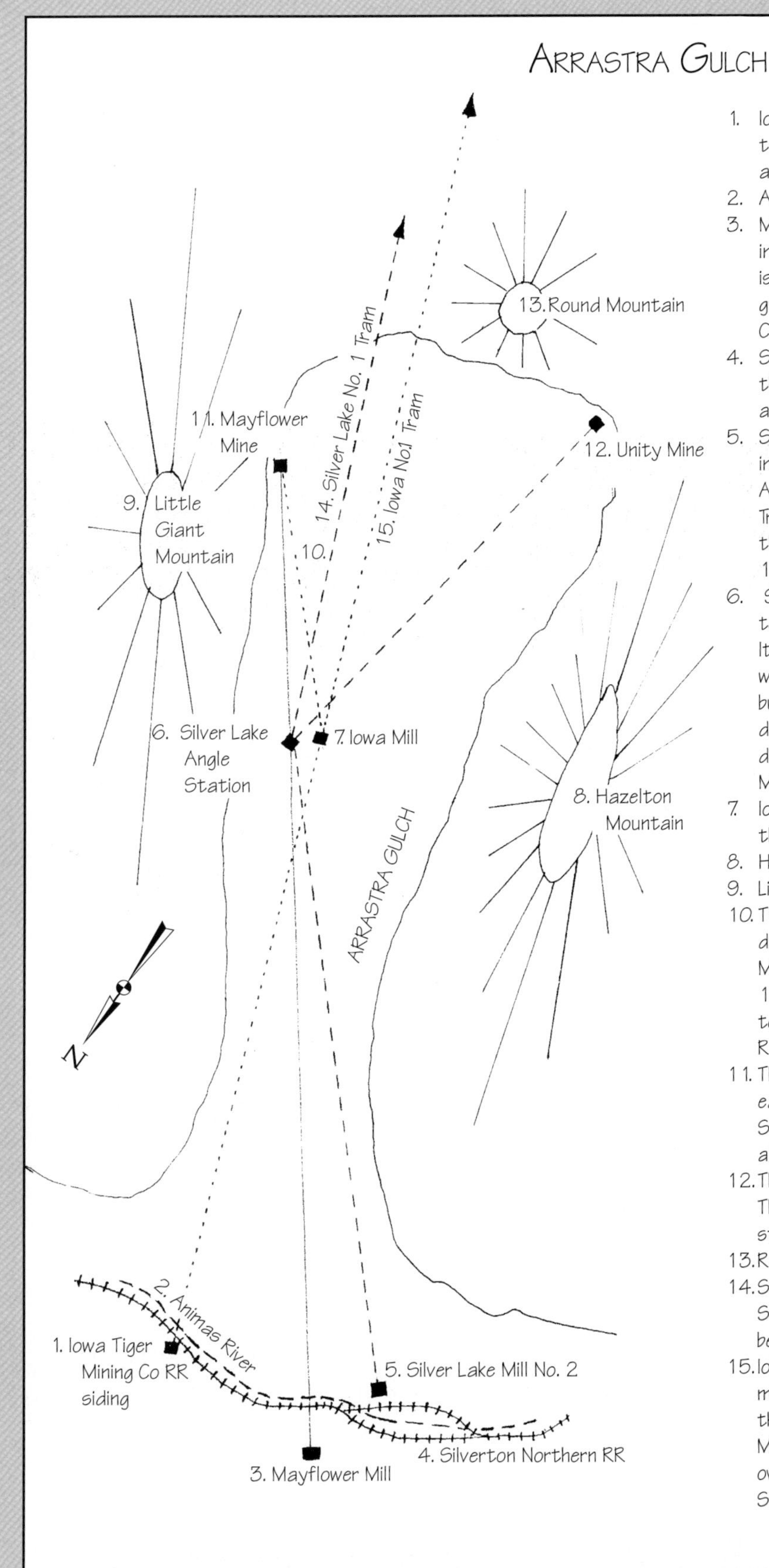

1. Iowa Tiger Mining Company railroad siding. The aerial tram was 6,000 feet long from here up to the Iowa Mill and was erected in 1898.
2. Animas River
3. Mayflower Mill (Shenandoah-Dives Mining Company) built in 1930. The aerial tram up Arrastra Gulch to the mine is 10,100 ft. long and is the only aerial tram in this gulch still existing and the only tram in San Juan County made with steel towers.
4. Silverton Northern Railroad, built in 1896, stretched up the Animas Canyon from Silverton 7 miles to Eureka, and eventually on to the town of Animas Forks.
5. Silver Lake Mill No. 2 was built in 1900. (No. 1 Mill , built in 1893, was situated at the end of Aerial Tram at Arrastra Lake, now known as Silver Lake.) The No. 2 Tram, starting here on the Animas River running up to the Angle Station, was 6,100 ft. long and was built in 1896.
6. Silver Lake Angle Station, built in 1895. This was the terminal for the No. 1 Tram coming down from the mine. It was also the terminal for the Unity Tunnel Tram which was part of the Silver Lake Mines. From here the buckets were transferred to the No. 2 Tram and sent down to the mill on the Animas River. This building was demolished in the late 1920's to make room for the Mayflower Tram.
7. Iowa Mill, built in 1895 and terminal for the tram from the Iowa Mine.
8. Hazelton Mountain.
9. Little Giant Mountain.
10. This 3,000 ft. aerial tram erected in the 1920's ran down from the Mayflower Mine to the Iowa Mill. The Iowa Mill ran the ore from the Mayflower Mine from 1928-1930 until the Shenandoah-Dives tram was completed to the Mayflower Mill located just above the Animas River.
11. The Mayflower Mine (Shenandoah-Dives), one of the early mines in Arrastra Gulch, became part of the Shenandoah-Dives Mining Company in the 1920's. It is also the upper terminal for the aerial tram.
12. The Unity Mine, part of the Silver Lake Mining Company. The tram was 4,000 ft. long and ran to the angle station.
13. Round Mountain.
14. Silver Lake No. 1 Tram, 8,500 ft. long, ran from the Silver Lake Mine in Arrastra Basin to the Angle Station below. This tram was completed in 1895.
15. Iowa No1 Tram was 9400 ft. long and ran from the mine in Arrastra Basin to the mill. Erected in 1895, this tram at its highest point at the base of Round Mountain Peak to the mill below had a vertical drop of over 2,200 feet in less than one mile. Both this and the Silver Lake Tram were considered engineering marvels.

Map created by Zeke Zanoni

The Arrastra area had once been one of the busiest areas in the entire state. Mining in the Silverton San Juans had started on Little Giant back in 1872. All the mines and trams indicated above had been operating around the turn of the century. But it would be the Shenandoah-Dives complex focused at the Mayflower Mine under the direction of Charles Chase that would keep mining and Silverton going into the 1950's.

Charles A. Chase in his office, 74 years young, 1950. Chase Family Collection. Courtesy William R. Jones

Charles A. Chase and the Shenandoah-Dives Mine

For over thirty years the story of mining in Silverton was primarily the story of one mine and one man: Charles A. Chase and the Shenandoah-Dives Mine. This mining operation helped Silverton survive the economic and social upheavals of depression and war which preserved it's historic character into the 1990's.

In the summer of 1925 a group of Kansas City capitalists hired mining engineer Charles A. Chase of Denver to come to the San Juans and find a new gold mine to buy and develop. Chase had made his reputation in Telluride as manager of the Liberty Bell Gold Mine. Unlike most engineers, his degree was in philosophy! He started his career as a surveyor and assayer, working his way up to General Manager. As he later wrote, "I, not ingrained with precedents of the industry, tried many fresh approaches." His methods won respect from laborers and capitalists alike. The Liberty Bell made a healthy profit on some of the lowest grade ore in the state.

Chase came to Silverton to examine the Old Hundred Gold Mine. While sampling it's lofty outcrops he could see another even larger vein a mile up the valley called the Shenandoah-Dives. Chase recommended against the Old Hundred, advice later operators of this mine wished they had followed. Instead he developed a bold plan to consolidate the Shenandoah-Dives, North Star, and Mayflower mines into one large mine twice the size of the old Liberty Bell.

The Shenandoah-Dives Mining Company finished construction at the Mayflower Mine portal in Arrastra Gulch in 1929. The mine featured a five story boarding house housing engineering offices, dormitory, commissary, kitchen, dining room, library, billiards, and even a radio! The ore was crushed underground and hauled in buckets down a two mile long aerial tramline to the mill. Here the ore was ground to a fine sand and the valuable gold and silver, along with the base metals, were separated from the worthless rock.

Just as construction was nearing completion the stock market crashed in New York and within months the prices of copper, lead, and silver declined so much as to be unprofitable. Only gold kept the mine afloat but as they tunneled deeper into the mountain the gold slowly diminished. By 1932 with nearly every mine in the state closed, Chase asked the men to voluntarily reduce wages by twenty-five percent. He also called on local merchants and landlords to lower prices and offered to mediate disputes over debts. He explained that only if full tonnage production could be maintained would the company and the town survive the "economic crisis" we now call the Great Depression.

The men and the townsfolk agreed, and Chase kept his word. No one was laid off and wages were restored two years later. Out of town creditors of the company were not so easy to convince. They threatened to force the firm into bankruptcy and liquidation. Chase's detailed financial analysis, eloquent letters, and persuasive arguments convinced the creditors to let the mine stay open. For years Silverton locals repeated the story of Chase telling creditors, "We could liquidate the mine's ore reserve and pay you off but it would destroy the mine and destroy the town of Silverton."

Chase made crucial efforts to save the narrow-gauge train, today the lifeblood of Silverton's tourist economy. In 1932 the train was blockaded for over 90 days. The D&RGW refused to open the line and instead wanted to abandon it. Chase literally snowshoed over the frozen tracks to Durango and traveled to the Interstate Commerce Commission in Washington, D.C. armed with photos of the poorly maintained line. On his return he brought back an ICC official who ordered the line opened and repaired to proper condition. Shipments from the Shenandoah-Dives kept the line going until it was discovered by tourists in the 1950's.

Until World War II brought an end to the Depression, Shenandoah-Dives with 150 employees was the largest single industrial payroll south of Leadville and west of Cripple Creek, an area covering 30% of the state of Colorado! Higher grade gold ore in the 1940's and higher prices due to World War II revived the company and the mining industry. Chase pioneered pollution control efforts finding water pollution "personally repugnant" 40 years ahead of his time.

While Chase's original mine plan and equipment were designed for low cost operation, Chase also knew the men were the foundation of the mine and took care of them. The boarding house was first class and the food excellent. A large free lunch and a pocket full of biscuits were given to every "tramp" miner who hiked the trail to the mine to ask for a job during the depression. Chase pioneered the contract mining system where men who were more productive earned more, yet lowered company costs. The men affectionately called him "Papa Chase."

Still, a bitter strike occurred in 1939 after the national Congress of Industrial Organizations (CIO) took over the local miner's union and decided confrontation could squeeze more out of "Papa." The dispute almost ended violently when fed up "home town" miners ran the CIO organizers out of town and later found weapons and explosives hidden in the union hall basement. Ironically Chase had been a pro-union man for years, unlike most of his mining contemporaries.

By the early 1950's times were changing. Chase, now in his seventies, still went underground every week and inspected the workings. But ore grades continued lower and the gold had long since been exhausted. Costs of labor and supplies rose steadily after the end of the war. Government policies changed to favor imported metals to help bolster anticommunist governments overseas. Finally, in November, 1952, Eisenhower and the Republicans were elected to replace President Truman (once a Kansas City politician himself.) Faced with Eisenhower's promise to end the Korean war, metals prices collapsed and with it the Shenandoah's future.

The company, marginal in the best of times, now began to lose money steadily. The stockholders had

In 1941, C.A. Chase, Francis Guire, C.L. Annett, mine superintendent, and Herb Mellus, mining engineer, sit on top, proud of their mine which had been producing for a dozen years and was destined to run another twelve. *William R. Jones Collection*

Charles Chase on his weekly mine inspection; the white suit and pressed pants a tradition he maintained into his seventies. *Zeke Zanoni Collection*

never received a penny in dividends over the mine's twenty-seven year life. Total profit for the company was a meager $1.2 million dollars and had all gone to pay debt interest and continued mine improvements. As the original stockholders sold out, control of the company passed to speculators and stock manipulators. Shenandoah stock which had soared during the Korean war also collapsed angering the new stockholders. Chase struggled with a new board of directors who knew little of mining and cared even less about a tiny mining town in the remote mountains of Colorado.

In February 1953 company directors voted to close the mine. Chase kept the mill running until March 20th to help out the smaller Silverton mines who depended on Shenandoah for ore processing. With Shenandoah's closing, these small mines also closed, mostly for good. Chase spent the last three years of his life in poor health but still fighting to reopen the mine and concerned as always with social and economic issues. His last official report warned that "... growing inflation is a growing threat." He was writing fifteen years before inflation became a major concern in American life.

Thus while the rest of the country enjoyed the economic boom of the 1950's, Silverton languished in a mini-depression. Ironically this lack of "modernization" preserved the historic Victorian look of the town enjoyed by tourists today.

Chase's legacy also survived. When Standard Metals Corporation reopened the Sunnyside Mine in 1959, it was staffed by miners, engineers and millmen who had all worked for Chase at Shenandoah. Chase's Mayflower Mill was used for processing Sunnyside ore until 1991. The old mill and tramline are now destined to become a museum and tourist attraction helping Silverton survive into the new century. "Papa" Chase would approve. —*Bill Jones*

Author of this article, Bill Jones, a friend of the Chase family and a collector of history of the Shenandoah-Dives Mine and others stands outside the Old Hundred Mine in 1994. *John Marshall Photo*

The Shenandoah-Dives Mine

nestled against the rocky cliffs of Little Giant Peak as it appeared during its operation from 1930 to 1953 under the direction of Charles A. Chase.

1. The tram, running through steel towers, delivers empty buckets back from the mill to the tram house.
2. Note the telephone lines running on top of the towers.
3. A building left from the original Mayflower mine was converted into the tram house. This is base level and led directly into the mine.
4. A catwalk gives outside access to the upper buildings.
5. The stub tram leaves from the back of the tram house and runs on wooden towers up to the stub tram house.
6. Stub tram house. Here supplies were ferried by overhead monorail to the boarding house or by rail to the timber shed.
7. The boarding house. The upper two stories, with the balconies, were for the miners' quarters. The floor showing at the bottom was the third floor and housed the commissary, among other things. The bottom two floors, not showing because of the terrain, were the main floor with dining facilities and the basement with the furnace and kitchen help's living quarters. Note the trail to Silver Lake stretching off in the background.
8. The timber shed with the overhead power lines. Inside the timber shed was the electric shop with the lamp room for the miners' lights. First and foremost though, it was the building covering the portal to the main level of the mine. There was a blacksmith shop just inside that portal.

Let's spend some time exploring the setup of this mine—how people lived and worked there.

The Shenandoah-Dives was a special mine, close to town, and closely associated with the town. Thanks to the skill of Charlie Chase, the mine kept running on slim margins through the depression and World War II and kept 190-200 men employed.

As with any big mine, the Shenandoah-Dives was a consolidation of claims running from the Old Shenandoah in Dives Gulch by Cunningham over to the Mayflower Mine in Arrastra Basin and covered all the previous claims in-between. The new access to these areas was through the underground expansion of the old Mayflower workings high on the left side of Arrastra Basin and was known collectively as the Shenandoah-Dives Mine. Ore from these workings was milled at the new Mayflower Mill which was built down by the road at the bottom of the new tram.

Lucy Baker used this picture as a postcard in 1945. She wrote on the back: "This is where I live. The pinhole (second window from the left, bottom row) marks my room window. There is one story below that you can't see in this picture. There are five stories all together." *Wilma Bingel Collection, Dot Bingel Photo*

Boarding Houses

were essential to many of the bigger mines. However, few are still standing today. It seemed extremely fortunate to be able to discover, fifty years later, pictures and stories of this boarding house from people who actually lived and worked there.

The large building that was the Shenandoah-Dives boarding house reached up five stories. It was a well-insulated building and the kitchen was fully electric. Usually, single men were made to live here, sharing a bedroom with one and sometimes two others. The building was coal heated and there was a big boiler downstairs so there was always hot water for showers. The engineering office was in the boarding house and so was the commissary. If you had a meal at the boarding house it was all you could eat, and waitresses served you! And it cost just sixty cents. The original Shenandoah boarding house was clear over by Cunningham Gulch. That mine ran from around 1902 to 1920. The boarding house there was hit by an avalanche in 1906 and twelve men were killed. They rebuilt that house in the same spot and put avalanche fence controls above it. The house stayed there until the forties when Charles Chase had Joe B. Salazar burn it down to keep Shenandoah-Dives mine employees from using it for their personal pleasures while they worked at the Spotted Pup.

Below, the men are in the process of building the Shenandoah boarding house. The year was about 1929.
Greg Leithauser Collection

The water for the new boarding house in Arrastra was piped out from deep in the mine around 3480 on the main level, back toward the "44" raise. An old shack was located there where a lot of water came out of the rock nearby. The shack was re-timbered to make a reservoir to collect the water. The water was pumped to a holding tank close to the shifter's office. To keep it from freezing the men diamond-drilled so the pipe ran by the dry room and supplied the wash basins and so on in there. Then it ran on into the boarding house through the rock, never having had the chance to freeze. There was an abundant supply of this good water. Fire hoses were kept on the dif-

Lucy Baker, the head cook, stands to the left of her daughter, Dot Bingel. Dot helped with the cooking and was also a waitress. Working high in a mountain valley but with more than a hundred men always around, Dot wasted little time finding four different husbands, however she had only one son. She was ahead of her time.

Wilma Bingel Collection

ferent levels of the boarding house just in case. Unlike so many buildings in the mountains around Silverton, this one never burned. Funny, some buildings were burned by accident, some to make room for new structures, and some by intention to remove them from county tax rolls. The Shenandoah-Dives boarding house instead was ignominiously torn down for materials in the sixties.

"When I first rode the tram I fell in love with it. I'll bet I got a million miles in riding on that darn thing. Up and down. It was wonderful. So was that boarding house. The top two floors were for us miners. We had two guys to a room, sometimes three. Single beds. When I worked there, seems like we had 150 men living there. You could go from the boarding house, to the dry room, to the main level. A guy never did have to go outside.

"And you know a man could eat anytime. I used to work with a guy who'd carry his thermos so it wouldn't take up room in his pie can and then he could fit five sandwiches into it along with pie and fruit. He even used to get us to try and take some sandwiches for him. God, he loved to eat.

"We had electricity. The building was coal heated and there was a big boiler downstairs so there was always hot water for showers. I used to get off of night shift and just sit there and listen to the old miners' stories. They'd go on and I loved hearing them talk."—*Bill Rhoades*

This picture, taken from a little higher up on the mountain than the one below, clearly shows how the monorail splits from the stub tram and branches off to run to the back of the boarding house. This is how coal was brought in to fuel the furnace, 1945.

Dot Bingel writes in 1945: "This is the boarding house where I work. The main floor is the dining room and kitchen, the two upper stories are rooms for the miners. We are feeding close to sixty men now." Wilma Bingel Collection

Fine Dining

Joe Vota turns and smiles for the camera in this rare glimpse of the Shenandoah dining hall in 1934 (he is the third man from the left in the above picture). Sure, the Swedes would sit with the Swedes and the Italians with the Italians. Racism? No, but made it a lot easier to have a conversation. The cooks, bakers, waiters, waitresses and a janitor lived just below on the bottom floor where the furnace was also located. The table in the foreground is where a savory plate of food would often be left out for the night shift. The kitchen, bakery, steam tables, stoves, walk-in refrigerator and supply room were all on the right side of the dining hall.

Herb Mellus Photo. Courtesy William R. Jones.

"Box Car Kelly and Clyde Pitts were wonderful storytellers. Hell, I was just a kid. I loved to listen to them talk. We'd get off night shift and go sit at the coffee table in the boarding house. Sometimes there'd be fresh doughnuts. Often, there was hot soup or dry cereal and coffee and these guys would sit and tell us stories. Their mining experiences and how to mine stories would float out on the late night air. Sometimes we'd be up all night. I couldn't get enough of listening to them."

—Bill Rhoades, 1994.

Wired—No Joke

"Oh, sure. There was always someone ready to play a joke on you. One time this assistant electrician wired the doorknob to our room in the boarding house. We came off shift and he was hidden where we couldn't see him. Every time we tried to open that darn door, he had this crank and he'd wind it up and shock the heck out of us. Well, we caught on and followed those wires. There were three of us and only one of him but we were all friends. We let him live." *—Bill Rhoades*

Above the mess hall on the third floor was the commissary. Also on that floor was the office of the mining engineer, Joe Vota, and the geologist, Herb Mellus, 1930.

William R. Jones Collection

The Commissary

And just what could you get in this packed commissary? Pretty darn near anything. Money was rarely exchanged. It was all put on a tab and taken off your paycheck. The concession was leased from the Shenandoah. In return, Stewart Edlund and Vince Tookey did the payroll for the men and took care of first aid. Everything came up on the tram. Every now and then even tailors would come up and measure the men for suits. Those measurements would be sent to Chicago and two weeks later the suit would come back in the mail and up the tram. Then the men would look quite elegant on their next trip to town. Yes, the mail came up on the tram and was given to the men in the commissary. Catalogues—possible competition—seemed to get burned rather than delivered, but no one complained because they didn't need money at the commissary for their purchases. And the variety of goods available seemed endless. Candy, magazines—no booze—but soft drinks, toiletries, medicines, mining gear, boots, hard hats, cotton underwear and socks, boot packs, T-shirts, oxfords, Pendleton shirts and blankets, leather jackets, jewelry cases, watches, Parker Pen sets, rings, women's nylons, undergarments, house dresses. What? Oh yes, people from town, the miners' wives, would also ride the tram up and go shopping. They found children's clothing, pants, dresses, and socks. More? Okay, radios, record players, records. There were fishing and hunting season and holiday seasons to stock for. At Christmas time, two tables would be brought up from the mess hall and electric trains running around the edges provided a cheerful diversion while the center of the tables were filled with toys for the miners to buy for their kids. And if all that didn't get your attention, there was a pool table, a snooker table and a poker table. My, what a place!

The famous punchboard game, rediscovered in Tommy Savich's wonderful museum in 1994. These boards kept miners busy and short of change in their spare time. John Marshall Photo

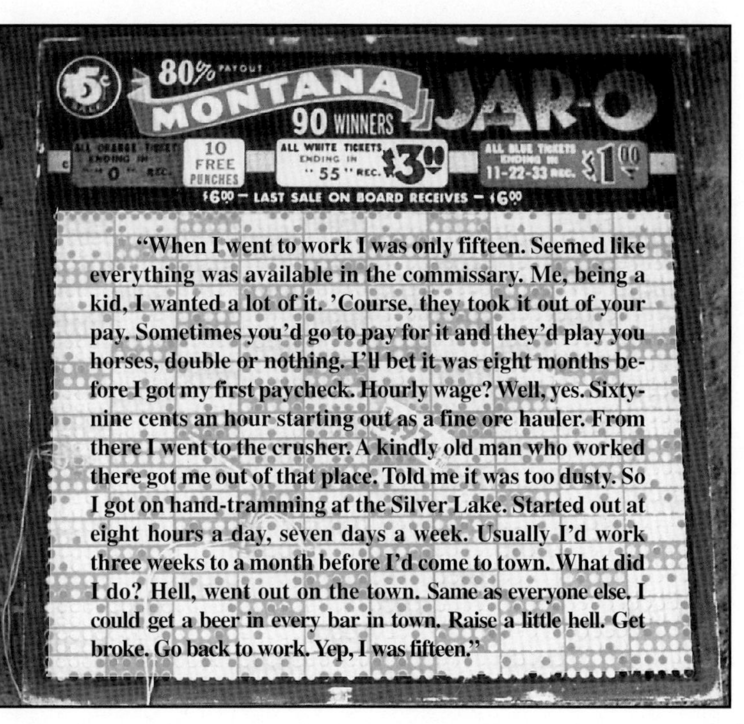

"When I went to work I was only fifteen. Seemed like everything was available in the commissary. Me, being a kid, I wanted a lot of it. 'Course, they took it out of your pay. Sometimes you'd go to pay for it and they'd play you horses, double or nothing. I'll bet it was eight months before I got my first paycheck. Hourly wage? Well, yes. Sixty-nine cents an hour starting out as a fine ore hauler. From there I went to the crusher. A kindly old man who worked there got me out of that place. Told me it was too dusty. So I got on hand-tramming at the Silver Lake. Started out at eight hours a day, seven days a week. Usually I'd work three weeks to a month before I'd come to town. What did I do? Hell, went out on the town. Same as everyone else. I could get a beer in every bar in town. Raise a little hell. Get broke. Go back to work. Yep, I was fifteen."

The commissary clerk, Vince Tookey (1943) worked for Stewart Edlund, who had the Packard agency in Silverton but no used cars. (He just used that to buy himself a new car every other year.)

Vince Tookey Collection

Shenandoah lodging: two bunks, a radiator, a window, usually a table and a chair and hot showers and flush toilets just down the hall. All this luxury in a building that was wrapped up in an old tram cable bolted to the rock walls out back in hopes of keeping the building where she stood. Herb Mellus Photo, William R. Jones Collection

Home to Bed

Ten to twelve hours. Another hard day. But at last you're getting off shift and you're one of the lucky ones—you live at the boarding house. No long tram ride in the dark and cold. No drive to town through deep new snow. Nope, just home to bed. You head for the adit at Main level, sloshing through water, the shadowy, damp dark walls dripping in the glow of your headlamp. An empty pie can hangs from your strong, work-hardened hands, you are ducked over, shoulders hunched. Once you stumble. Now ladders to climb. Maybe take a hoist. At last, the familiar work areas near the portal. There was the shifter's office, and right across from it was the powder room. The shifter kept the key to the magazine. Right by the powder room was the access to the lower levels and the crushers: a telsmith cone crusher and a Symons. Below that was the -250 Level or Base Level where the locomotives would load their cars. There were three or four chutes and they usually pulled two cars of ore out to the tram house where the crushed ore was loaded onto the tram and shipped down to the mill. It was quite unusual for a mine to crush the ore before sending it down and it was one of the unique features of the Shenandoah. Past the shifter's office, turn left and pass the compressor which was the heart of the mine. Next would be the blacksmith shop where there was access to the timber shed containing the electric shop and the lamp room where all the miners' lights were kept. And the dry room (like a locker room) was just around the corner. There were showers there, too, and in the dry room a man could clean up before he rode the tram down the hill to town. Since you lived at the boarding house you could go from the dry room through a tunnel, mostly under the protecting rock, past the First Aid room and into the main level of the boarding house. A snack, a cup of coffee, some conversation and off to bed.

Of course bed wasn't always a great place to be. Dick used to be a jockey and drank vinegar to keep his weight down. And kept a gun under his pillow. One night when Dick and his buddy Jim went to town for a little fun a rock came through the roof and smashed into his bed. This time his drinking was probably a good habit. ❋

Dick Riggna and Jim Bergamaschi stand on the stub tram tracks in front of the timber shed. Rich Perino Photo

Leach Zanoni is second from the left along with three women who are about to go underground. They all have carbide lamps. Check out the old hard hats on the right (early 1930's). They were made out of treated leather and were darn hard. The new hats were called Hard Boiled hats; that was their brand name and from then on hard hats were hard boiled and so usually were the men who wore them. *Zeke Zanoni Collection*

Leach Zanoni on the left, Zeke's father, and Vern Hamish, mine superintendent, on right, 1935. Vern was the first mine boss to allow women to come underground. He was a heck of a nice guy and that's probably why he got away with it, because it sure bothered a lot of the old miners. In the past, if a woman came into the mine, they would all walk out.

Zeke Zanoni Collection

CONNIE TREGILLUS REMEMBERS her childhood visit to the Shenandoah workings: "It was 1937 and three of my male cousins were here visiting from Philadelphia. My father was A.P. Root, longtime Silverton assayer. Father decided we should go for an adventure and ride the tram up to the Shenandoah and see a working mine. I guess his assaying background and his being a stockholder in the company was some help. (Riding the tram was not usually a pleasure trip.) Even so, I remember we all had to sign permits before we could board the buckets. Oh boy! That tram took right off across the canyon and over the Animas River. I was only fifteen and I was scared. We had a couple of pieces of wood we put across the bucket to use as seats, but the earth seemed so very far below us. As we got up Arrastra, the ground kept getting closer and I got more confident. I remember we saw a red fox on the way up. And lunch was waiting for us at the boarding house. There were lots of men and we had curried chicken gizzards with rice. It was wonderful. I think the men ate pretty darn well. Then, we all put on carbide lights and followed Father into the mine. There, just before the crushers was something called a Winze—I think. We rode that down to the -400 level where the Swedish Forest was. It was a great big open room with round timbers called stulls placed randomly around the cavern. That's why it looked like a forest. A lot of gold was found in here and was a great asset to the mine. We had to keep our eye out for engine-driven ore cars and would stand aside as they went by. All too soon it was time to go, but, you know, that was some fifty years ago, and I can see it all today, clear as a bell."

A.P. Root and Mr. Norton standing outside their assay office. It was located on 14th Street, across from the Wyman Hotel by today's Kendall Mountain Cafe, 1907. *Cynthia Francisco, Connie Tregillus Collection*

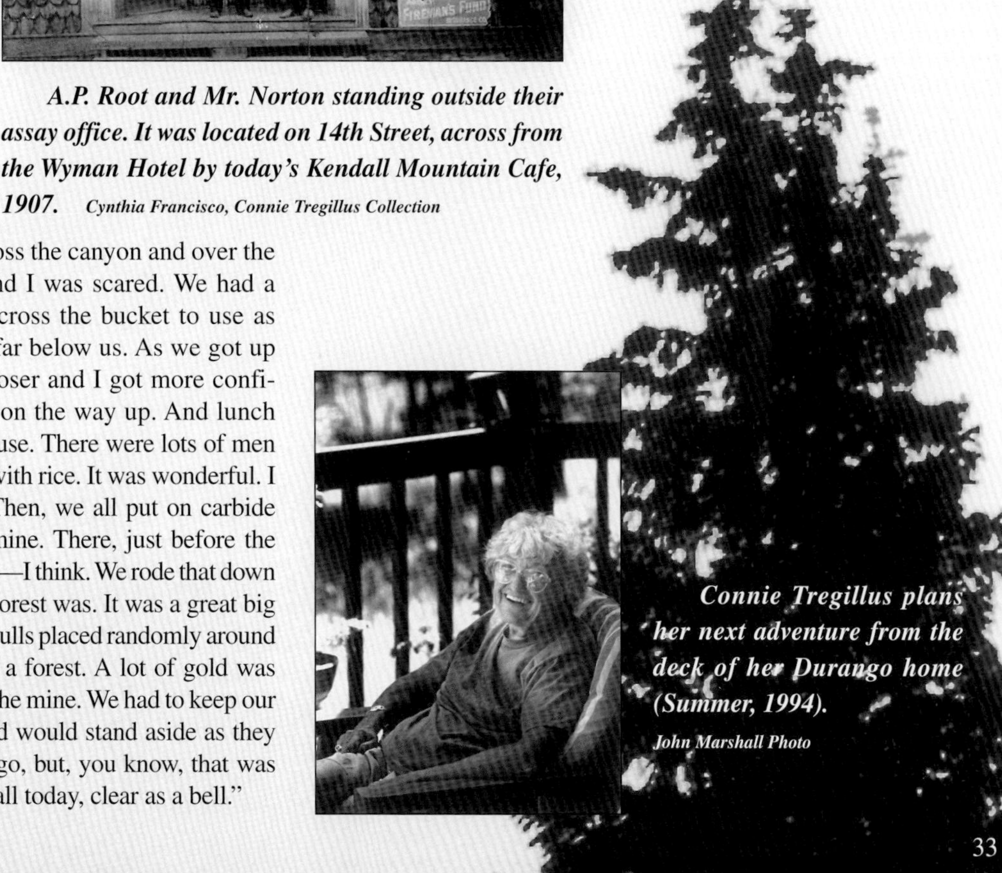

Connie Tregillus plans her next adventure from the deck of her Durango home (Summer, 1994).

John Marshall Photo

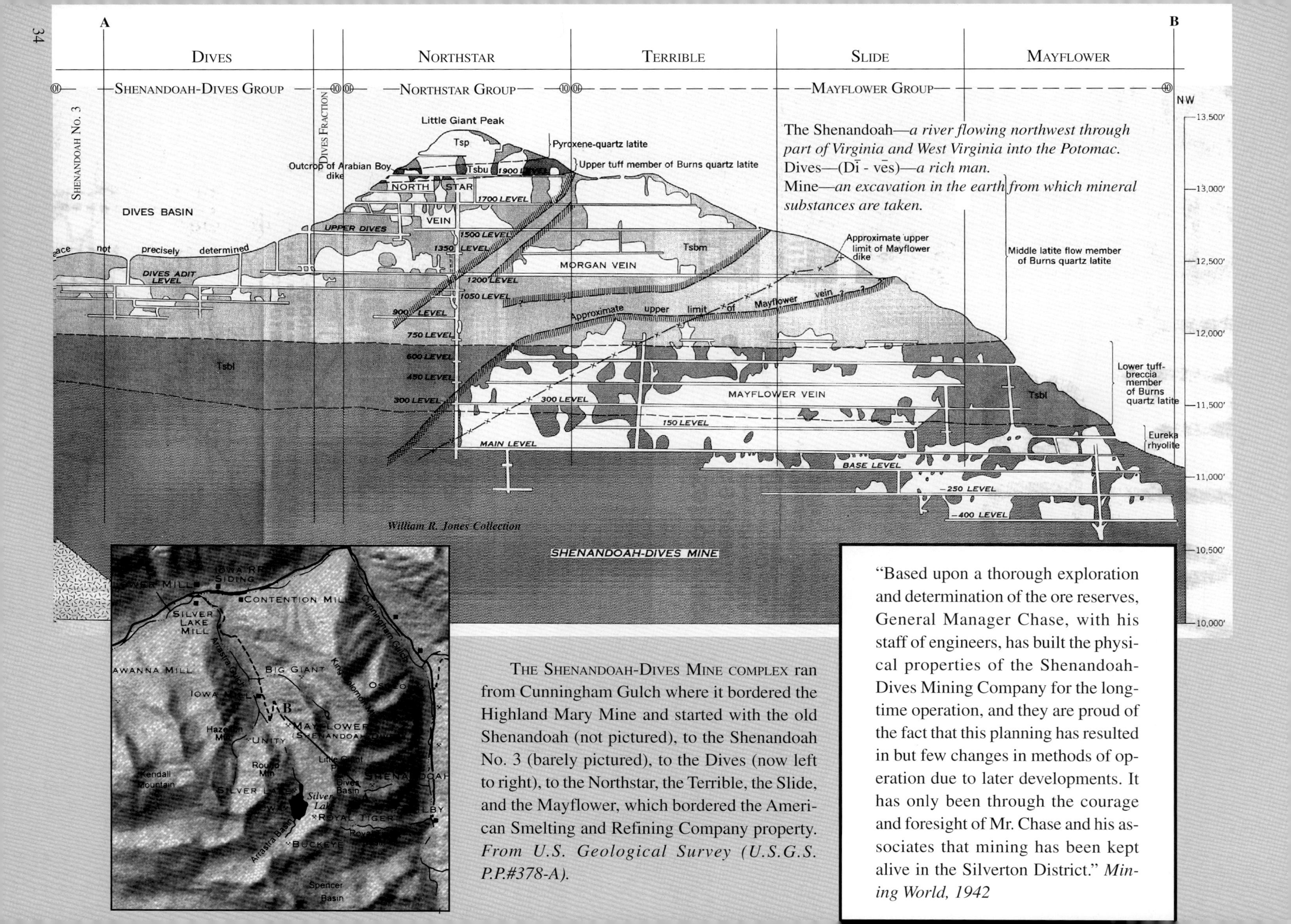

William R. Jones Collection

The Shenandoah—*a river flowing northwest through part of Virginia and West Virginia into the Potomac.*
Dives—(Dī - vēs)—*a rich man.*
Mine—*an excavation in the earth from which mineral substances are taken.*

THE SHENANDOAH-DIVES MINE COMPLEX ran from Cunningham Gulch where it bordered the Highland Mary Mine and started with the old Shenandoah (not pictured), to the Shenandoah No. 3 (barely pictured), to the Dives (now left to right), to the Northstar, the Terrible, the Slide, and the Mayflower, which bordered the American Smelting and Refining Company property. *From U.S. Geological Survey (U.S.G.S. P.P.#378-A).*

"Based upon a thorough exploration and determination of the ore reserves, General Manager Chase, with his staff of engineers, has built the physical properties of the Shenandoah-Dives Mining Company for the long-time operation, and they are proud of the fact that this planning has resulted in but few changes in methods of operation due to later developments. It has only been through the courage and foresight of Mr. Chase and his associates that mining has been kept alive in the Silverton District." *Mining World, 1942*

Vince Tookey Photo, 1941

Mining World, April, 1942

If you lived in town and it was wintertime and you'd just ridden up the tram, you'd enter the mine on this, the -250 level or base level. This is the level where the crushed ore is loaded and taken to the tram. If lights turned red that meant an electric locomotive pulling a string of ore cars was coming and you better stand aside, because they didn't stop for anyone. Now you have two ladders to climb, past the two crushers to reach Main Level.

If you came from the boarding house you would already be on Main Level and you would pass the compressor and might even see Carl Bleich, the chief mine mechanic standing beside it.

Heading Underground

SEEMS LIKE IT'S TIME TO HEAD UNDERGROUND. The Shenandoah-Dives Mine appeared over the years in various publications. In 1932, *National Geographic* discussed it and displayed pictures of it in an article on Colorado. In 1942, a magazine called *Mining World* featured the mine on its front cover and ran a second article on it in June of that year. Some of the technical aspects included here come from these issues. Most important to the mine was the compressor, the heart of the operation.

"Air for the thirty working machines, drills, etc. in average daily operation is supplied by a 2200 cubic foot Chicago Pneumatic simplatic valve compressor, driven by a 250 h.p., 440-v., synchronous motor."

And in the time that compressor ran—from 1929 to 1953 and again from 1959 to 1961—it only burned out the motor once. That was around 1947 and Howard Hill and John Glanville, electricians in the shop, surveyed the damage. The best and fastest way to fix it, they advised Charlie Chase, would be to get a company electrician out of Denver. The man and his helper showed up and with John and Howard and Joe Salazar and Joe Todeschi helping, the compressor and the mine were back to running in a little over twenty-four hours.

Joe Todeschi did much of the maintenance on the machine. Every six weeks the oil was changed, the brass bearings were tightened and the valves adjusted. The efforts paid off, for without the compressor the mine didn't run.

Standing near the mine portal with the compressor showing on the right. William R. Jones Collection

Drilling and Blasting

The work of a drift miner.

If you're going to mine in the San Juans, you better be ready to do some drilling and then some blasting. The rock here is extremely hard. Most of it is latite and high in manganese and quartz. So, if you're going to drive some drift, first you've got to make a hole in the rock so you can place your dynamite—your powder. And of course, the idea is to drive drift on the vein where the gold is going to be found, not in the waste rock where all you find is added expenses. It doesn't always happen that way, though.

In the old days, those holes were drilled by hand and it was called single jacking or double jacking. In 1907, in the Old Hundred Mine, you were given a four-pound hammer (see opposite page) and a drill steel (see page 42). That hammer was driven with one hand and by the end of the day it must have felt like it weighed a hundred pounds. Now, the drill steels you are hammering can vary in lengths, and when they get dull you change to a sharper one. The principle of single jacking was to hit the steel and then turn it an eighth of a turn. This would keep the hole nice and round. A good man could drill about twelve inches in a thirty minute period—all things being equal and ideal—which they never were. This would go on all day, ten hours a shift. In 1907 you could make three bucks a day and at that time it was darn good money. A lot of guys would bring their own hammer, passing on the company's generously provided one, just because they knew the feel of their own and liked it better. Of course, if you broke it, it was at your own expense. And sure, a man could live in the boarding house at the mine—three meals a day, hot showers and a bunk— but then you only made two dollars a day.

Now the double jack was a little different. It required a two-man drilling team and careful teamwork. One person was the hammer man and the other was the steel tender. You would take your double jack, which was a long-handled hammer weighing six, sometimes eight, pounds and swing down on that steel. Hard! The principle was the same as the single jack, but now your partner held the steel while you swung away. And then you'd switch off. There wasn't any room for error if you wanted to keep your body and your partner's intact. Hold that steel! Straight, steady and true. Don't forget to rotate it. Guys could hit that steel forty to sixty times a minute. (In contests or mining celebrations some men could muster seventy-two strokes a minute, but not for long.) The true secret of hand drilling was to hit that steel square and to make sure the rotation was correct. But it only counted if the drill was good—sharpened and tempered properly. That's why a good blacksmith was essential to the operation.

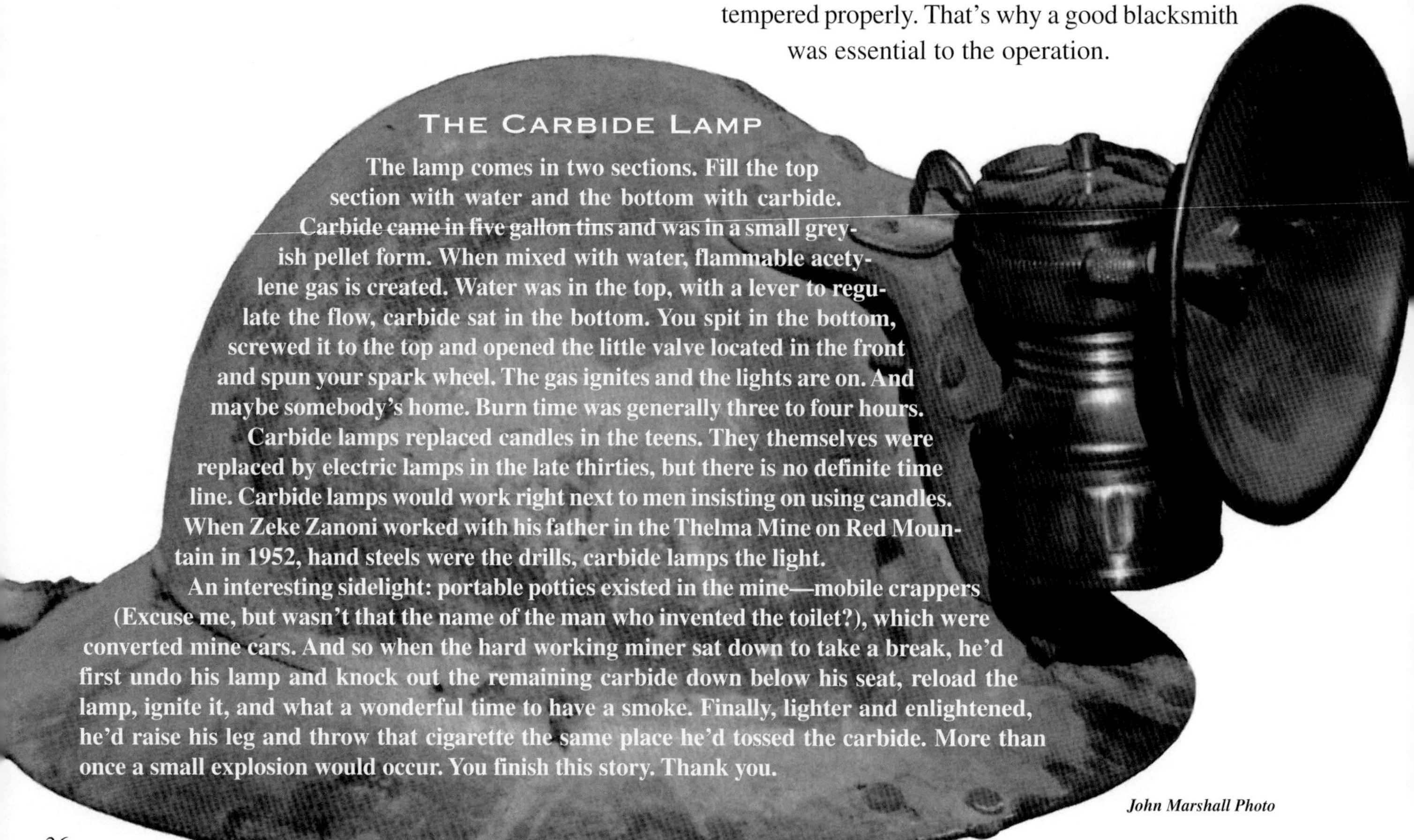

The Carbide Lamp

The lamp comes in two sections. Fill the top section with water and the bottom with carbide. Carbide came in five gallon tins and was in a small greyish pellet form. When mixed with water, flammable acetylene gas is created. Water was in the top, with a lever to regulate the flow, carbide sat in the bottom. You spit in the bottom, screwed it to the top and opened the little valve located in the front and spun your spark wheel. The gas ignites and the lights are on. And maybe somebody's home. Burn time was generally three to four hours.

Carbide lamps replaced candles in the teens. They themselves were replaced by electric lamps in the late thirties, but there is no definite time line. Carbide lamps would work right next to men insisting on using candles. When Zeke Zanoni worked with his father in the Thelma Mine on Red Mountain in 1952, hand steels were the drills, carbide lamps the light.

An interesting sidelight: portable potties existed in the mine—mobile crappers (Excuse me, but wasn't that the name of the man who invented the toilet?), which were converted mine cars. And so when the hard working miner sat down to take a break, he'd first undo his lamp and knock out the remaining carbide down below his seat, reload the lamp, ignite it, and what a wonderful time to have a smoke. Finally, lighter and enlightened, he'd raise his leg and throw that cigarette the same place he'd tossed the carbide. More than once a small explosion would occur. You finish this story. Thank you.

John Marshall Photo

Positions of the men would be changed every seven to ten minutes. This procedure should give you about a thirty-inch hole in roughly thirty minutes. Those holes were drilled in varying patterns, loaded with powder, fired, and with luck and skill the rock was broken. Progress was slow (approximately thirty inches advancement per round), but that was the way it was done in the San Juans in the hand steel era.

All things changed with the coming of the air compressor. Most of the gear underground became pneumatic—run by compressed air. In the early days, those compressors were driven by steam engines often located on the surface right next to the compressor they were designed to run. Coal and wood were the fuel. Today, the compressors are usually underground and run with electric motors and they also distribute air throughout the mine in air lines or pipes hung off the ribs of the tunnels and drifts. The invention of air compressors created the air drills. Those first drills were 'dry' drills with no water to retard the dust. They quickly became known as widow-makers. The rock dust created by the drilling went right into the lungs. Silicosis was the result. Miners con or consumption. Rock in the box. Life expectancy was about forty years.

Today, wet drills prevail. In most mines, water is collected in a sump and pumped into pipes that run throughout the mine along with the air lines to the working locations. The water comes through a hose into the drill. It travels down the water needle in the drill to be forced through the hollowed out center hole of the drill steel. This eliminated ninety percent of the dust and also cooled the bit as it was drilling and washed the cuttings from the deepening powder hole.

The end result of drilling and blasting is to create a drift. A drift is usually eight feet by eight feet. The top is the back, the sides are the ribs, and the bottom is the floor. Terms like the footwall and hanging wall are used to describe the walls of a vein. Now for this particular round, we're going to drill twenty-seven holes. This number will vary from round to round but for now on page forty we show a diagram using a twenty-seven hole round. Each hole is drilled about two feet apart and six feet deep. When all the holes are drilled, it's time to move the equipment out of the area because you're going to start loading them up with powder. All the holes are loaded virtually the same way. The sticks of powder to be loaded in each hole are split with a knife before being placed. This allows the powder to be tamped in tightly against the rib of the hole. The split first stick is called the primer stick. The cap or detonator is inserted in that primer stick with the fuse crimped to the detonator.

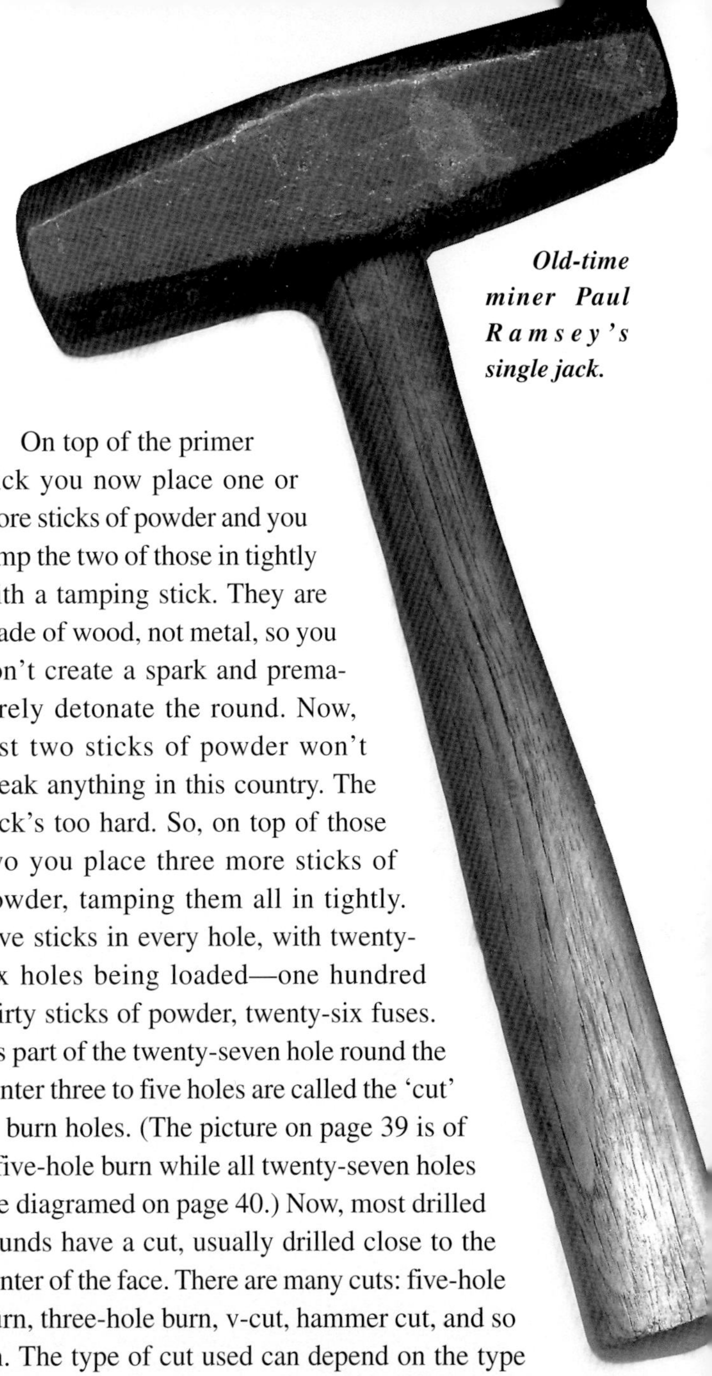

Old-time miner Paul Ramsey's single jack.

On top of the primer stick you now place one or more sticks of powder and you tamp the two of those in tightly with a tamping stick. They are made of wood, not metal, so you don't create a spark and prematurely detonate the round. Now, just two sticks of powder won't break anything in this country. The rock's too hard. So, on top of those two you place three more sticks of powder, tamping them all in tightly. Five sticks in every hole, with twenty-six holes being loaded—one hundred thirty sticks of powder, twenty-six fuses. As part of the twenty-seven hole round the center three to five holes are called the 'cut' or burn holes. (The picture on page 39 is of a five-hole burn while all twenty-seven holes are diagramed on page 40.) Now, most drilled rounds have a cut, usually drilled close to the center of the face. There are many cuts: five-hole burn, three-hole burn, v-cut, hammer cut, and so on. The type of cut used can depend on the type of ground, whether one is stoping, sinking a shaft or driving drift—among many other factors. This example shows a 'drift round,' drilled in the heading or a face. The 'modern' miner using a 'jack-leg' drill usually puts in a five-hole burn cut. For this round, that's what we'll use. (Just remember there are lots of variations. A whole book's worth of variations have been recorded in *The Blasters Handbook.)*

Fuses are cut nine or ten feet long, leaving more than two feet of fuse dangling out of every hole. When it comes time to light the fuses, and that, folks, is called 'spitting' the fuse, there are at least two critical factors that come into play—the length of each fuse and the sequence in which they are lit. Timing is everything. If the holes don't explode completely and in the proper sequence then a lot of very dangerous things can occur. Many a man has been killed

"The mining of the ore bodies is done by the conventional shrinkage method in the Shenandoah-Dives. In both the stopes and the drifts, 5° ft. powder holes are the general practice and are filled with 45% du Pont Gelex. In dry ground, blasting is with No. 6 du Pont Caps and National Fuse. In wet ground, du Pont Electric Caps, fired by 440 v. current, detonate the powder charge. In drilling the powder holes, Timken detachable bits are used, starting at 2" with five changes and finishing 1°". The drill for the stoping machines, which are about evenly divided between the Gardner-Denver and Ingersoll-Rand is 1". On the leyners (Gardner-Denver 89's and Ingersoll-Rand's 35's) 1$\frac{1}{8}$" steel is used." *Mining World, 1942.*

Ernie Kuhlman demonstrates the drilling process with a pneumatic stoper. Old Hundred Mine Tour, 1994. *John Marshall Photo*

Gene Luther, above, and Zeke Zanoni, below, are headed to the powder magazine in the Shenandoah, 1959. The shifter's doghouse is to the right in the bottom photo. *Zeke Zanoni Photos*

These pictures were taken after the C.A. Chase era of the Shenandoah-Dives had ended (by 1953); Standard Uranium, which soon became Standard Metals, had reopened the mine in 1959. This is the powder magazine. It was located across from the shifter's office. The shifter would keep the key to the magazine. *Zeke Zanoni Photo*

or maimed by drilling into an unexploded stick (or more) of powder. When a round is done properly it will result in a lot of rock being broken. When you're deep in the mine and around the corner, hopefully safely out of harm's way, the resulting explosion will sound like a series of muffled booms, not just one report.

You can't get halfway through spitting your fuses using your trusty Bic and run out of fuel. Never mind the fuel, there's a lot of air and water moving around underground to help that light go out. You have to have something you can rely on so you make what's called a spitter fuse. You cut about two feet of fuse off the fuse reel. It has powder all down the center of it. With your pocket knife, you notch that fuse about every inch, halfway through to the powder core. Then, when you light one end of the spitter fuse, it will burn down the powder center and spit out a very hot flame at each notch, like a Fourth of July sparkler. The spitter fuse provides perfect timing and it won't go out. The old timers even used a spitter made of lead tubing whose core was powder.

In drilling and blasting with dynamite or powder, you must consider the line of least resistance through the rock. If you don't then the shot can result in what is called a *freeze.* The rock will fracture but it won't move. A bad situation.

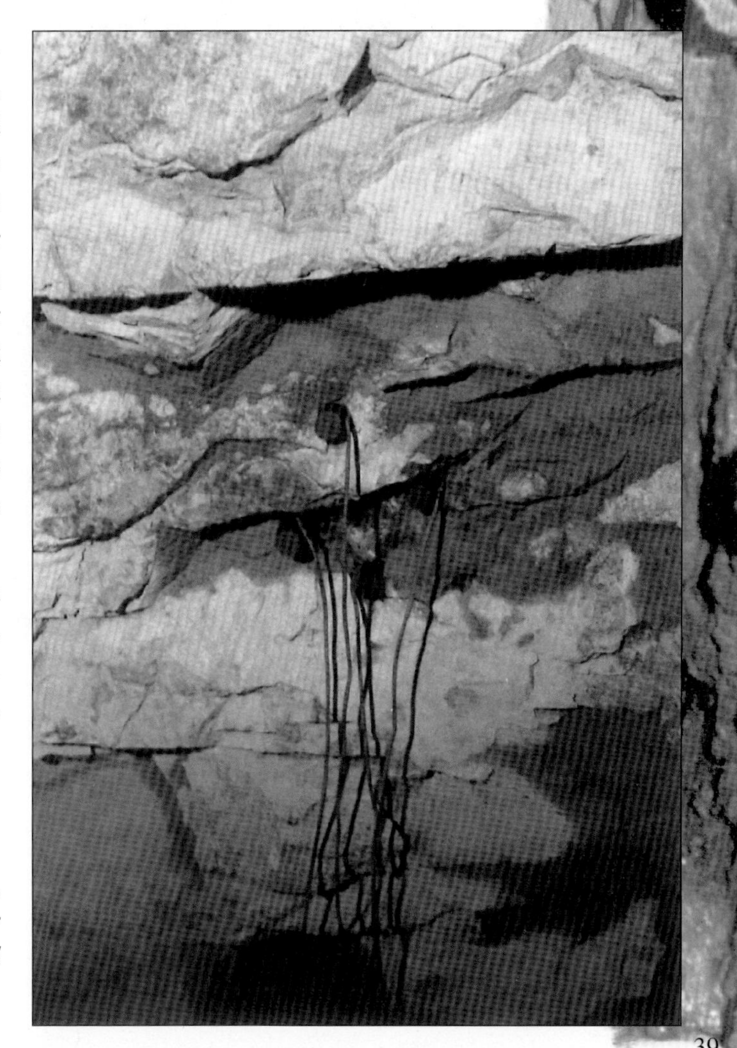

The burn holes, in this case a five-hole burn, on a face waiting for ignition. The empty center hole is there but partially hidden under a small rock overhang. *John Marshall Photo*

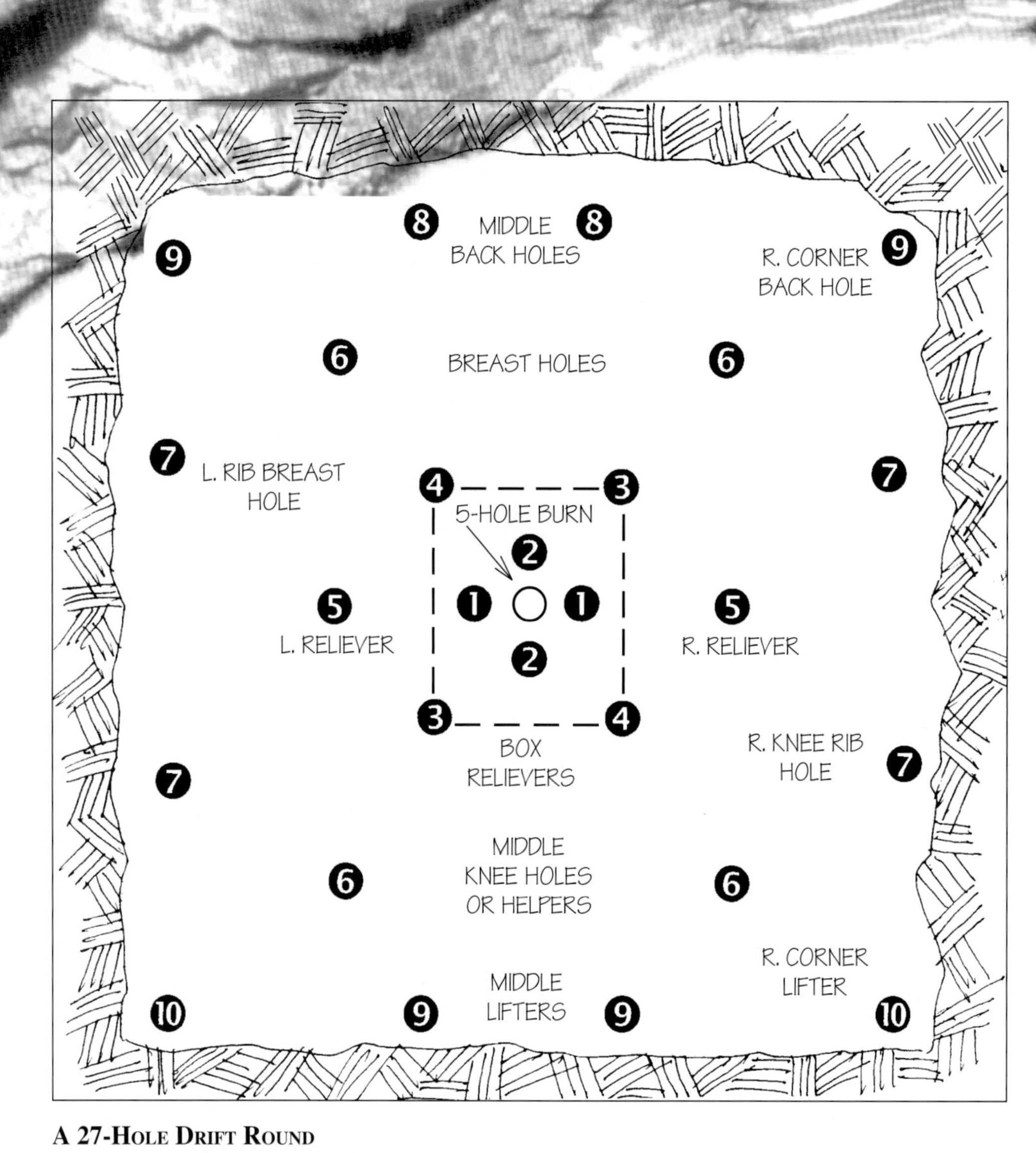

A 27-Hole Drift Round

The number of holes drilled in a round depend upon the size of the round, hardness of the ground, the type of cut used and the geological formation of the rock.

The number in the circle ❶ indicates the firing sequence of that particular hole. Some of the holes can be timed differently as shown above, but the cut holes must be fired first and the lifters shot last as shown in the diagram.

Every hole plays an important part in breaking its share of rock. But, pulling the cut (the five-hole burn in this example) is critical to the success of the round. If it does not pull to its full depth, the rest of the holes will "bootleg" breaking only to the depth of the cut. This situation proves costly and dangerous to the miner. Great care is always taken in drilling the cut.

That's why there's nothing in the center hole. That empty hole is for expansion. Now remember, what you light first, detonates first, second, second, and so on. The cut is timed to the second: the burn, ❶ and ❷ , are fired first and second, then the box relievers, ❸ and ❹ , are fired. Since the tied-together burn holes detonate first, the rock will break to the line of least resistance, to the center hole. And thus, you've extracted the hole core from the face—six feet deep. The next ring that blows is called relievers, and it will break into the core of the burn. The third ring that goes off is called trim or black holes. The last ones are called the lifters. They literally lift all that broken rock right up from the hole and outside the face so you can clean it up and start all over again.

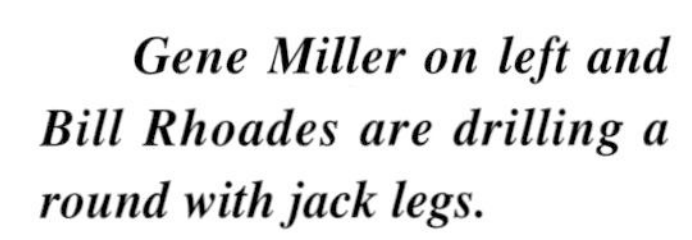

Gene Miller on left and Bill Rhoades are drilling a round with jack legs.

Zeke Zanoni Photo

Now that the round has been drilled and shot, the two drift miners know their work is only half done. The other half will be removing the broken muck (rock) the following morning and advancing the rail (railroad track). But first they must scale down all loose rock still hanging to the back and ribs with bars.

An eight foot by eight foot round, six feet deep, will produce around thirty-eight tons of broken rock. In the last fifty years this rock in a tracked drift will normally have been picked up with an Eimco Overshot Mucker and put into three-ton cars to be hauled off and dumped, taking about three hours. Two tie holes will have been hand dug and within them are placed new railroad ties and then the *slide rail* is pushed forward six feet and temporarily spiked down. On every third round, the drift will advance enough for the slide rail to be spiked into permanent position. Along with this will come the advancement of the air and water as well as the ventilation line. If all goes well, this entire process of mucking (the previous day's blast), drilling and blasting will take place in one eight-hour shift, and is called *'cycling a round.'*

Wally Grey on left and John Dillon are mucking up a drift round that had been shot the night before on Main Level of the Sunnyside.

Zeke Zanoni Photo

A timber crew builds a San Juan chute in the Sunnyside shortly after the turn of the century.

Zeke Zanoni Collection

The sweat would flow and at times blood would show, but the work went on. For this is contract mining in the San Juans and the miner was paid for per-foot advancement. Thus the work of the drift miner. Only the best of the crusty veterans would prevail. Would you like to give it a try?

The drift miners are but one of several teams used in the complex of a producing mine. There are the shaft, raise, prep and last but not least, the stope crews. It is the stope miner who produces the large quantities of ore needed to make the mine a profitable operation, each with his own story and mining technique. But, what of the support groups—the timberman, blacksmith, trammer, mechanic, electrician, car whacker, nipper and the high wire boys? The list goes on and on, and without them the mine would come to a standstill. How vital they are! ❋

MY FATHER, LEACH, had started work up at the Thelma. That's a claim up by the Silver Ledge above Chattanooga and seen below the highway. The Thelma became the last patented mine in San Juan County. I'd been following my father around since I was seven or eight while he checked out a lot of the old claims in the area, looking for silver or gold. By the time I was eleven he'd stuck a shovel in my hand and my mining days were starting.

The Thelma was the first real underground job I had. My father and Ernie Hoffman were partners on this one. It was around 1952 and I was fifteen. We were using carbide lamps and hand steeling. The ore was good and was getting milled down at the Mayflower Mill until they closed and we were forced to quit. I was the gopher and the hand trammer and I would load and push those one-ton cars to the outside dump, among other things. I can still remember watching my first

Highblaster

Nippers and powder

(top left) A highblaster prepares a powder charge on the end of a blasting stick. This stick is fifteen to twenty feet long and will be used to reach and bring down high hangups of rock.

(top right) Nipping powder at the bottom of the Washington Shaft on the American level. The powder will get loaded onto the mancage for distribution through the mine.

(right) A typical scene on the F Level mechanic's shop.

(bottom right) Loading timber, three by twelves, to be hoisted by the Washington Incline Shaft.

Underground in the Sunnyside around 1982.

Zeke Zanoni Photos

Mechanic's shop

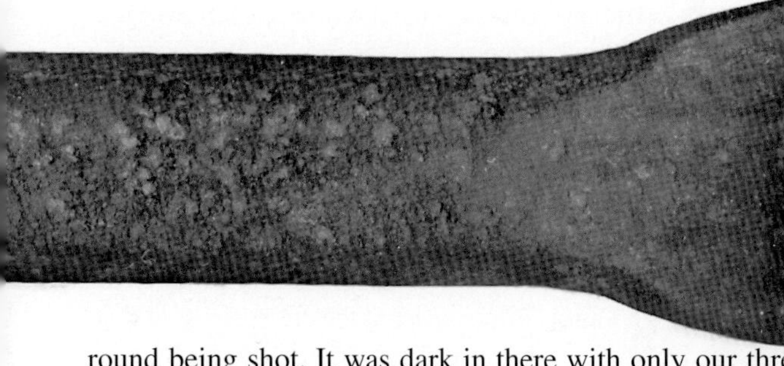

round being shot. It was dark in there with only our three carbide lamps. Fuse after fuse was being spit and it was getting awfully smoky in there. I did what any sane man (or young kid) would do and headed for the portal at high speed. I was just getting going when a big hand landed on my shoulder. "You stand right there until they're all lit, son." I was learning. This mining was in my blood.—*Zeke Zanoni. His grandfather was a miner in Silverton. And his father, and his brother too. And so is Zeke. "My children? Nope, no miners there. That's O.K."* ❈

Material movement

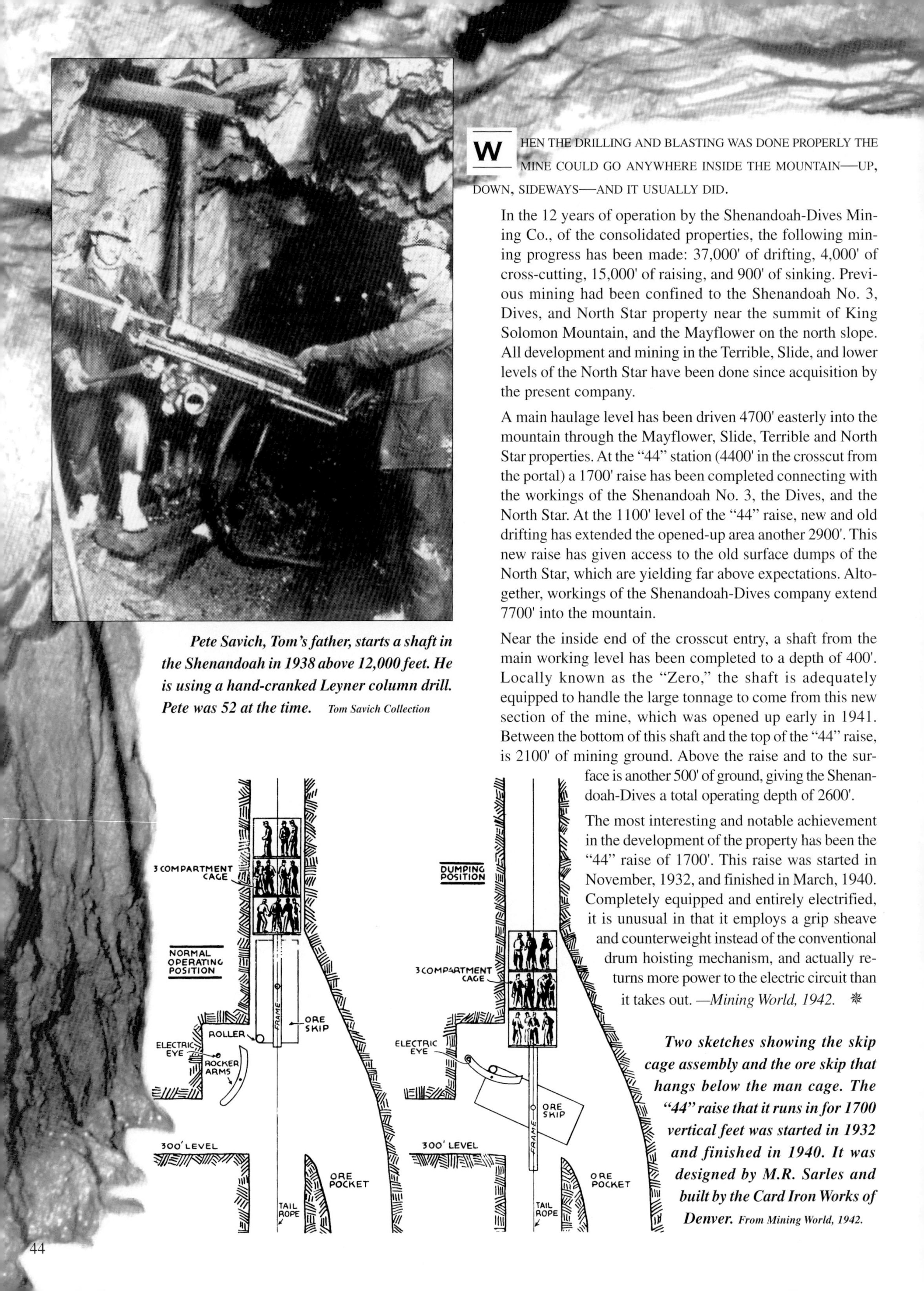

Pete Savich, Tom's father, starts a shaft in the Shenandoah in 1938 above 12,000 feet. He is using a hand-cranked Leyner column drill. Pete was 52 at the time. *Tom Savich Collection*

WHEN THE DRILLING AND BLASTING WAS DONE PROPERLY THE MINE COULD GO ANYWHERE INSIDE THE MOUNTAIN—UP, DOWN, SIDEWAYS—AND IT USUALLY DID.

In the 12 years of operation by the Shenandoah-Dives Mining Co., of the consolidated properties, the following mining progress has been made: 37,000' of drifting, 4,000' of cross-cutting, 15,000' of raising, and 900' of sinking. Previous mining had been confined to the Shenandoah No. 3, Dives, and North Star property near the summit of King Solomon Mountain, and the Mayflower on the north slope. All development and mining in the Terrible, Slide, and lower levels of the North Star have been done since acquisition by the present company.

A main haulage level has been driven 4700' easterly into the mountain through the Mayflower, Slide, Terrible and North Star properties. At the "44" station (4400' in the crosscut from the portal) a 1700' raise has been completed connecting with the workings of the Shenandoah No. 3, the Dives, and the North Star. At the 1100' level of the "44" raise, new and old drifting has extended the opened-up area another 2900'. This new raise has given access to the old surface dumps of the North Star, which are yielding far above expectations. Altogether, workings of the Shenandoah-Dives company extend 7700' into the mountain.

Near the inside end of the crosscut entry, a shaft from the main working level has been completed to a depth of 400'. Locally known as the "Zero," the shaft is adequately equipped to handle the large tonnage to come from this new section of the mine, which was opened up early in 1941. Between the bottom of this shaft and the top of the "44" raise, is 2100' of mining ground. Above the raise and to the surface is another 500' of ground, giving the Shenandoah-Dives a total operating depth of 2600'.

The most interesting and notable achievement in the development of the property has been the "44" raise of 1700'. This raise was started in November, 1932, and finished in March, 1940. Completely equipped and entirely electrified, it is unusual in that it employs a grip sheave and counterweight instead of the conventional drum hoisting mechanism, and actually returns more power to the electric circuit than it takes out. —*Mining World, 1942.* ❋

Two sketches showing the skip cage assembly and the ore skip that hangs below the man cage. The "44" raise that it runs in for 1700 vertical feet was started in 1932 and finished in 1940. It was designed by M.R. Sarles and built by the Card Iron Works of Denver. *From Mining World, 1942.*

COYNE THOMPSON SHOWED UP AT THE MINE WITH A DEGREE IN JOURNALISM AND HISTORY. Did "Papa" Chase, who had a degree in philosophy, suspect that here was a kindred soul? In any event, Coyne was hired and worked there in 1938, 1939 and 1940.

"I had just graduated from Colorado University. It was 1938 and jobs were scarce. I knew Silverton well and decided, 'What the heck, I'll go stand in line for a job at the mine' I didn't think I stood much of a chance for a job but Mr. Chase hired me. Right away I rode up on the tram. It was a twenty minute ride and then the work began. My job was mucking out the rails, keeping the rocks off of them. I had a partner and we roomed together in the boarding house. I remember I was always hungry and always ready to sleep. Maybe it was the hard work, maybe it was the high altitude, but when dessert time came around there was always pie to eat. I'd get off shift at 4:30 and dinner was at six. Serving was family style, and you helped yourself from the bowls and platters that were passed around. I can still remember when the pie came out, especially the mince pie. I'd have, not a piece, but the whole darn thing. Mmmm, good! Then I'd go upstairs to my room and read for a while, maybe a book I'd brought up from the library in town, maybe a magazine I'd purchased from the commissary. Some of the men would stay up there for months at a time. We could go through the mine, right up to the top, and from there we could look out of the portal and see Silver Lake. Sometimes we'd walk down to her and come back to the boarding house on the trail.

Coyne Thompson at home in Durango in 1994. Skiing is still one of his loves, and in 1993, he and five of his friends did some figuring as they started down the trail: the oldest in their group was 85, the youngest was 76, and together they had over 400 years skiing experience. John Marshall Photo

For a young man the work was hard but hardly ever boring. One time we were working in a raise and somebody had cut a water tank in half and rigged it so it could be pulled up these tracks to get us to where we were working. We used a signal to get back down below which was a bucket hung on the end of a wire. One pull meant for the man below to start the contraption and bring us on down. It got to be time to come off shift, and my partner and I got on that damn half water tank, yanked the bucket and started down. It was a nice smooth ride all the way down, right to the bottom 600 feet below. We climbed out, stood up and looked around. Nobody there. Nowhere. Whoever had been there last had somehow set the brake just hard enough to let us down. Not much safety there, but we were lucky. Tommy Knockers? There were a lot of things happening that defied explanation. Care for some mince pie?

Bill Rhoades in town in 1994. *John Marshall Photo*

Sure, I had some close calls. Didn't everybody? Got buried once. We'd been going into where an old stope was supposed to be. We drilled some short holes. My partner went back to get some powder. That's when the roof caved in. He came back and found a little piece of blue showing between rocks. That was my jacket. He got me out of there. I had a broken jaw, a broken nose, and cuts on my head that were going to take eighty stitches when we could find the doc. I swore I'd never go back in a mine again. So as soon as I healed up, what did I do? Went right back underground, of course.

It seems there was no shortage of ingenuity in the men who worked in these mines. It didn't matter whether they were under ground or down the hill in the shop. Practical, clever, even brilliant. Here they made a 9˘ ton locomotive, managed to haul it up the hill two miles and then get it into the mine ready to work. These were the men that could.

The haulage system on the main level consists of three-ton Granby type cars built by the Card Iron Works, operating on a 30" gauge track of 45-lb. rail and moved by trolley locomotives. One of the locomotives operating on the main level is a 6-ton Goodman, while the other is a 9˘-ton locomotive built in the company's own shops under the direction of Mr. Sarles and Howard Hill, chief electrician. This locomotive is equipped with two 40-hp. Westinghouse d.c. motors and hydraulic brakes. A dead-man control is built into the seat so the motors cannot be operated unless the operator is seated. To aid in servicing, the controller, which is mounted in a horizontal position together with the other electrical equipment, is contained in a compartment on top of the locomotive frame and can be removed as a unit by a crane. This gives quick and easy access to the running gear. Another unusual feature is the incorporation of hydraulic jacks in the end plates. They are operated from controls set in the cockpit. Should the locomotive become derailed, one man can get it back on the tracks by lowering the jacks and placing irons under the wheels.

Three 3-ton locomotives are also available for use in any of the workings.

On the 250' sub level, which connects with the mine terminal of the aerial tram, a three-car train of 5-ton V-bottom Card cars hauls the crushed ore approximately 500' to the tram bins. To save room, and to release the locomotive for other work, one of the ore cars was recently motorized with a 10-hp. Westinghouse 250-v. motor. This motor, together with the controls and trolley, was mounted on the rear end of one car and connected to the wheels with a chain drive. *Mining World,* 1942. ❋

Bill Rhoades started out in the mine in 1942 driving locomotives. This is out by the timber shed in 1945. *Wilma Bingel Collection*

Down in the valley bottom one crosses the Animas River and starts into Arrastra Gulch. Here (at right) we are looking up towards Little Giant Peak from the Animas River. The mining has stopped but the road, the tram and the power line still lead up the hill. Off to the right of the peak lies Silver Lake (shown below) in the heart of Arrastra Basin. Early on, the majority of the mining was centered there, with the Silver Lake Mine, Iowa, and Royal Tiger all producing ore. With the coming of the Shenandoah-Dives in the late 1920's, the activity shifted down into Arrastra Gulch below the Silver Lake Falls.

John Marshall Photo

THE SILVER LAKE DISTRICT OF ARRASTRA BASIN

Up in Arrastra Basin lies Silver Lake and the Silver Lake complex. The mill is behind the men and the boarding house section is to the right. The men in the boat? Maybe early members of the Silver Lake Yacht Club. They weren't fishing. The mill was built 1893-95, so the picture is probably right around 1900. *Zeke Zanoni Collection*

SILVER LAKE

When the old workings of the Silver Lake Mine were reached via expansion underground, there were high hopes of finding valuable ore, either left in the old workings or by new exploration.

> Southwest of the Shenandoah-Dives property on King Solomon Mountain is the Silver Lake group, one of the earliest producers of the Silverton district. This mine is owned by the American Smelting & Refining Co. and has been idle for a number of years. To reach the property, that company contracted with the Shenandoah-Dives Mining Co. to drive a 4200' crosscut tunnel from about 1100' in on its main haulage level to a point underneath the old workings of the Silver Lake, and also to handle all transportation from the workings to the portal and over the tram. The increased traffic from the Silver Lake development over the tracks of the main haulage level of the Shenandoah made it advisable to install an automatic block system (a signalling system for the trains, activating red lights ahead down the track to eliminate head-on collisions) from the portal to the Silver Lake crosscut. —*Mining World,* 1942.

The ore trains from Silver Lake would join the tracks of the Shenandoah on the main level. There the cars would be dumped into the two crushers near the portal by the compressor and the timber shed. From the crushers the ore would be carried out on the lower -250 level to the tram, loaded in, and brought on down to the mill.

In 1959, six years after the Shenandoah-Dives closed, Standard Uranium, which later became Standard Metals, reopened the mine. They headed back into the Silver Lake cross cut, and tried to rework that area. This was at the same time they were busy driving the American tunnel at Gladstone. Their Silver Lake effort lasted about two years. Interestingly, about twenty years later this same outfit would work under another lake, Lake Emma, with dire results.

This is a good view of the steam-heated bunkhouse at Silver Lake along with the other buildings of the Silver Lake Mine, 1947. *Rich Perino Photo*

This is the view from the top of the Shenandoah looking down into Silver Lake Basin. The buildings in the foreground, right, are part of the Silver Lake Mine. In back on the right is the Iowa Mine. Across the lake on the left is the Royal Tiger. Wilma Bingel Collection

"George got me to climb these ladders. I swear they went on forever and I hate heights. George was always taking me to these most wonderful places. 'Just don't look up and don't look down,' he'd tell me. It seemed like we climbed forever. 'C'mon, we're almost there,' I'd hear him say. When I finally did get to the top I could go outside and look right down at Silver Lake. It was so special. The grass was all green and the wildflowers were blooming. Such beauty. I didn't want to leave. But, finally, it was time to go. Climbing down was easier. I figured I'd just get to the bottom and never go again. But, you know, if George had asked me, I guess I would have. It was that pretty." — *Wilma Bingel*

Zeke Zanoni Photo

Wilma and George Bingel at their 45th wedding anniversary in 1986 in Silverton. George worked at the Shenandoah from 1940 to 1945. Wilma Bingel Collection

Louis Dalla and Fenrick Sutherland talk over old times at Mary Swanson's 75th birthday party in the French Bakery in 1977. Gerald Swanson Collection

The Spotted Pup Ore Dump

The Shenandoah managers knew they had a valuable ore dump sitting in Dives Basin above Cunningham Gulch. It was a 60,000 ton, hand-sorted, high-grade ore dump made by the original Shenandoah Mine during its operation from 1902 to 1920. These mining companies were often owned by business consortiums located far away, often in Chicago, New York, or even England. Sometimes for no obvious local reason, these companies would suddenly abandon their properties leaving valuable equipment and ore behind. The Spotted Pup boardinghouse was still standing by the portal where the Spotted Pup ore dump was located. The year was 1949. The question was, how to get the ore, which would later prove to be surprisingly high in gold and silver. First, using the new Shenandoah-Dives access, a drift was driven on the 900 Level to come out beside the dump. Then the miners drove a 200-foot raise to the dump so there would be a place to shove the ore down in order to transport it over to the tram. That would require a cat and a cat skinner. Getting the cat skinner would be no problem. Louie Dalla was the man. Getting the cat to the 12,500-foot-high dump would be a little more difficult, but anything was possible. They took a used D-4 and proceeded to dismantle it clear back down at the Mayflower Mill. They had to cut the frame apart so it would fit on the tram. Up the pieces went, all the way to the 900 Level. Joe Todeschi welded it back together up there and Louie drove it to the dump. For at least two summers he worked pushing that ore down the raise. This Shenandoah silver kept the mine running during this time , and by 1951 most of that ore had been shoved down and trammed out. The old cat had worked to the bottom of the dump which proved to be a large patch of ice. Louie, while working hard one day, slipped on that

The crushed, rusting skeleton of the D-4 cat still lies below the Spotted Pup ore dump where it fell years ago. Zeke Zanoni Photo

The D-4 cat is working on the Spotted Pup dump in 1949. This photo is looking southeast toward Cunningham Gulch. To the right is the ridge between Dives Basin and the Royal Tiger Basin. *William R. Jones Collection*

ice and as the cat was heading over the edge, he quickly jumped off and landed on a pile of rocks. He carried a scar on his head for the rest of his days, proving how fate had spared his life.

Or did it? Apparently, insurance paid for the cat. And there was an awful lot of effort saved by eliminating the need to get a worn out cat dismantled, transported through the mine, down the tram and then back together again. Now, I don't suppose it was possible that Mr. Dalla just waved good-bye to that little D-4 as she trundled over the edge? Do you remember if he had a scar when he started this job?

Either way it doesn't matter, Jack Caine climbed up and salvaged a bunch of parts from that machine with a mule train later on in the fifties. ❋

Louis Dalla runs the cat on the dump 1949. When he left, the dump was gone and the cat was gone, but the stories linger on. *William R. Jones Collection*

Here, tram boss Joe Arietta stands on the first tower below the upper tram house, 1942. Note the recently-built avalanche deflector down hill on the right side. There's a story coming up about that. Rich Perino Collection

The Aerial Tram of the Shenandoah-Dives

So far we've explored the boarding house, the operation of the mine and some of the countryside around the mine. A large mine such as the Shenandoah-Dives demanded an aerial tram to provide a means of movement for the massive amounts of material required to operate the mine and carry the mineral-rich ore to the mill. Smaller mines depended on burros and mules or horses and wagons and sleds. In the valleys, the narrow gauge trains moved the most tonnage. In the mountainous country of the San Juans it was the trams that ran night and day through almost any kind of weather carrying large amounts of supplies. To be sure, the four-legged creatures provided assistance from time to time.

The first tram appeared, appropriately enough, in Arrastra at the Little Giant Mine in the summer of 1873. From that first 1,000-foot tram there followed at least fifty more trams of all shapes and sizes. At least a dozen of these fifty trams serviced the larger mines. Of all those, the Shenandoah tram proved to be the only tramway built with steel towers. It stopped a short distance below the boarding house at the base level of the Mayflower Mine. The tram house there contained a large bull wheel which turned the cable and led the buckets back down the hill. At the back of the main tram house was located one end of the small stub tram. It ran the short distance along wooden towers close to the ground up to the main level of the mine, the timber shed and the boarding house.

> "In 1929, when the present Shenandoah-Dives Company was formed, one of the first development projects was the construction of the 10,100' continuous bucket tramway, which drops in elevation from 11,200' at the mine, to 9700' at the mill site. This tram was designed by Fred C. Carstarphen of Denver. The

towers of the tram were built by the Pittsburgh Engineering Co., while the Stearns-Roger Manufacturing Co. of Denver fabricated the balance of the tram. The original installation was under the direction of Algot F. Andrean, who has continued since that time as superintendent. The rope speed is 500' a minute. Two sizes of buckets, 17 cu. ft., and 21 cu. ft., are spaced at intervals of approximately 400'. New track cable being installed is 1˘" lock coil rope made by the American Steel & Wire Co. The traction cable is a Roebling 7/8" 6x7 Langlay plow steel line.

Power for control of the tramway, normally actuated by the descending ore, is applied through two 6' grip sheaves at the upper terminal. These sheaves are of the pattern used in the "44" raise and previously described. A 50-hp. constant speed motor is connected to one of the grip sheaves. To the other sheave is connected a second 50-hp. G.E. motor of the variable speed type, which is used for reversing the tramway operation. With the buckets loaded with ore going down to the mill, power flows back into the line, and the motors are used only to control the speed and for power when the load is uphill.

Special four-wheel trucks and grips on the ore buckets were designed by Mr. Carstarphen. The tramway is Timken bearing equipped throughout trucks and cable supporting sheaves alike. The steel tramway towers also serve as support for the telephone communication system between the mine and the mill. Three interlocking telephone systems provide communication between points inside the mine; between mine, mill, and office, and also connect with the Silverton exchange of the Mountain States Telephone and Telegraph Co. At each tower is a signal switch and pull rope placed in such a position that tramway passengers may signal the tram operator in case of emergencies as the bucket passes the tower. Telephones also provide communication to the tram operator from maintenance crews working on the towers and cables.

At the mine terminal, a 300' auxiliary tram delivers loads of freight for the mine and hotel arriving over the main line at the main portal. In addition to Mr. Andrean, the superintendent, the tram system employs 11 men for operation and maintenance. —*Mining World,* 1942. ❋

Heading down. *Zeke Zanoni Photo*

The picture below is looking out from inside the tram house toward that first tower shown on the opposite page. Besides some of the towers, the tram house is the last vestige of the Shenandoah-Dives operation standing in Arrastra today. The tram house was originally used as a storage space or possibly even as a boarding house during the old Mayflower operation. When the Shenandoah took over in the late twenties, the building was converted into the tram house, and in the 1960's, when the Shenandoah boarding house was torn down for its materials, the old tram house was spared. If you squint into the sunlight here, you can see a load of timber coming into the tram house carried by timber hangers. Overhead and just to the right, out of the picture, are two chutes where the tram buckets are loaded with ore already crushed in the mine and hauled to the tram's ore bin by the locomotives previously described. This crushing of the ore inside the mine was one of the unique operations of the Shenandoah-Dives. If you were going to ride the tram down, you'd get on right here where Vince Tookey snapped this picture in 1941.

A spectacular view from the upper stub tram house. Note rails for ore cars which were pushed or pulled by a locomotive and the overhead monorail for the tram buckets which were pushed into place by hand, 1941. Wilma Bingel Collection

(left) The upper back part of the tram house is shown at the left. Wooden towers lead up to the upper stub tram house, the boarding house, and the timber shed, 1944. Rich Perino Photo

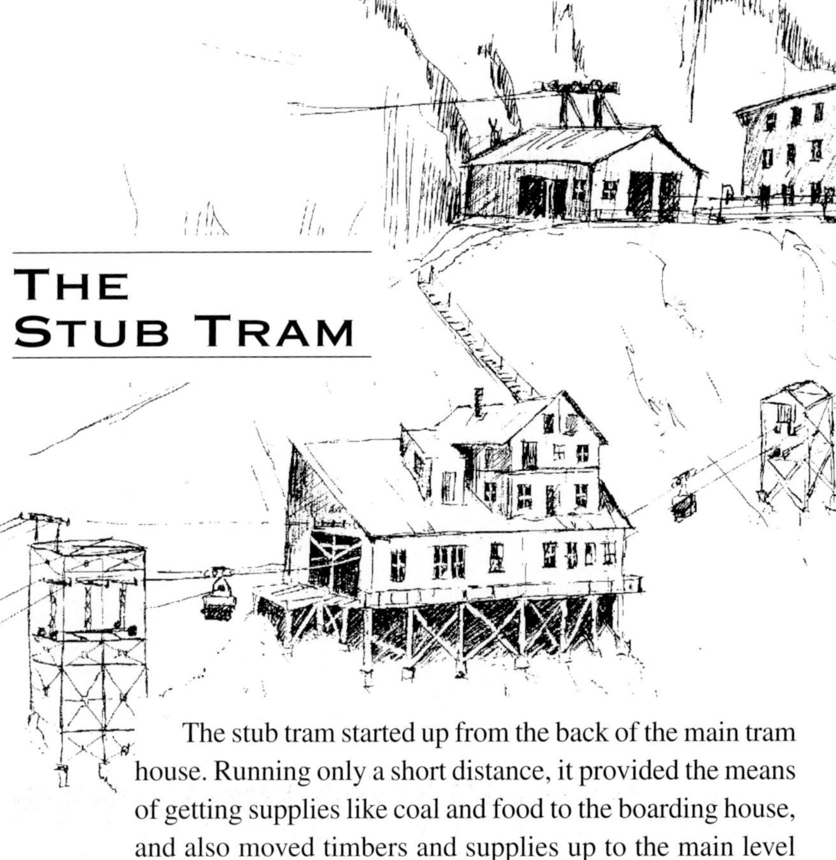

The Stub Tram

The stub tram started up from the back of the main tram house. Running only a short distance, it provided the means of getting supplies like coal and food to the boarding house, and also moved timbers and supplies up to the main level of the mine.

Another view from the upper stub tram house. The first tower out and the wooden walkway are the same as seen in the upper right picture, 1941. Wilma Bingel Collection

A view of the monorail leading from the stub tram house into the timber shed, 1942. Between the timber shed and the boarding house the ground would often be covered with materials waiting to be taken into the mine. The timber shed burned down in 1980. Lightning may have struck it, more likely, matches were the cause. **Wilma Bingel Collection**

This photo shows the timber shed in the center with the boarding house behind it. Beside that is the upper stub tram house with the tram towers running down to the main tram house. The Main Level tunnel entered the mountain directly from the timber shed, behind where the men are standing, 1946. **Richard Perino Photo**

"Bill Smith had a temper and, for him, work was work and fun was something you did when you wasn't working. So whenever the boys in the tram house would give Bill repeated bells for him to bring the bucket up and there wasn't any bucket, he'd get aggravated. If they kept it up, he'd get mad, and then he'd run down the hill with an axe in his hand. The boys would lock the doors, and it would be a while before anyone would signal the stub tram again."

Bill Smith worked the stub tram. He's shown here in the fifties, shortly after the mine closed. He had previously worked for Charles Chase at the Liberty Bell Mine in Telluride. **Zeke Zanoni Collection**

In town, just two short miles from the mill, there were kids busy growing up. Look at these two. Tell me they wouldn't be interested in a good adventure. The stories of what was going on in the countryside around them were already reaching their ears, maybe causing them to start making some plans of how to get in trouble a little later on? Jean and Gerald Swanson, goat-cart riders in 1934. *Gerald Swanson Collection*

The tram bucket hangs quietly in 1994, its forward motion forever frozen. *John Marshall Photo*

Kids and Trams Oil and Water

Gerald Swanson, Jim Drobnick and Richard Maes were together in junior high school in 1942. A special summer adventure for the boys would be to hike up from town to Arrastra to the Shenandoah-Dives Mine. Throughout the depression and over the years, out of work miners would take this same hike hoping to rustle a job. Even if there was no work they always would be fed, and might even enjoy a hot shower. This was the custom and these facts were known far and wide. Such was the goodness of mine manager Charles Chase to the men.

Well the boys knew this too and would conveniently time their arrival to catch the noontime meal. After the miners had finished, Mrs. Carey (Dot Bingel) would serve the boys some sandwiches and always some pie. Dot's pies had a regional reputation!

Afterwards the trio would help Bill Smith at the stub tram, loading and unloading supplies for the commissary and the boarding house; things like candy, punch boards, clothing, radios and clocks, sometimes jewelry and, of course, all kinds of good food. At the commissary there was a pool table and a snooker table and the boys might shoot a couple of games. Maybe they'd enjoy a cold pop from the boozeless bar.

The first tower from the mill stretching 128 feet above the ground, 1994. John Marshall Photo

The afternoon would pass too quickly for the boys and much more slowly the shift would end for the miners. Usually the single guys lived at the boarding house but the married men who lived in town would let the boys jump in the buckets with them for the ride down to the mill, the road, and the way back home.

The miners would make the boys duck down and hide from John and Joe Arietta, on duty at the bottom of the tram. The kids would jump out smartly as the bucket came round the bull wheel and run out the side of the building, usually with John or Joe's cuss words ringing in their ears. And in a couple weeks it would be time to go again.

Now, one day those young but experienced tram riders were walking the road back to town when a Ford station wagon pulled up next to them. Mr. Chase, himself, was driving it. "Hop in, boys. I'll give you a ride." It wasn't like he was asking, so in they went. Mr. Chase drove ever so slowly, expounding all the while on a particular problem that was occurring at the mine: "You know, young men, there are some people riding our tram who have no business doing so. They've been doing it a lot over the past couple of summers."

Gerald, Jim and Richard snuck furtive glances at each other.

"Now it's not right. It's against all regulations. And it's dangerous. If these people were to be foolish enough to keep it up, they will get into serious trouble."

No direct accusations. No names. Just a long, entirely one-sided conversation. The boys were relieved to see the drugstore come into view.

"We'll get off here, Mr. Chase. Thank you, sir, for the ride," they said as they jumped out and started to shut the door, but Mr. Chase had leaned over and was keeping it open. He looked solemnly at each one of the boys. "You do understand what I mean boys, don't you?"

We nodded our heads vigorously, not knowing what to say. We didn't ride the tram again that summer. And next year we only rode it maybe a couple of times. And that was it.

By 1953 Jean Swanson was still getting exciting rides. Don Robinson, her new husband, carries her past the Pickle Barrel in a wheel barrow. This uncommon custom was continued, and a ball and chain were added to the groom's foot. Appropriately, the wheel-barrowing begins at the courthouse and ends at the reception. Gerald Swanson Collection

The first tower out from the mine with the second one standing just below it—at least for a while, 1937. *Zeke Zanoni Collection*

IN THE SUMMER THE TRAM WAS USUALLY A GREAT WAY TO GET TO WORK AND OFTEN FUNNY THINGS HAPPENED ALONG THE WAY.

Going Bananas

On one occasion food supplies were being shipped up from the mill to the boarding house at the mine. A stalk of bananas broke loose from its heavy cloth sack before being put into a tram bucket. With fresh fruit considered a rare treat and with the opportunity presenting itself, the mill tram crew as well as several other mill workers helped themselves to the delicacy.

Finally on its way to the mine, the partially depleted stalk of bananas was spied by the "High Wire Crew" working on one of the towers. After they stopped the bucket and helped themselves to the treat, it was allowed again to continue on to the mine. The mine tram crew, seeing the almost bare stalk and knowing that they would probably be blamed for the missing fruit anyway, indulged in several bananas themselves.

When the boarding house manager realized the situation and saw no more than five bananas left on the stalk, he immediately summoned the mine superintendent. Vern Hamish, looking over the practically empty banana stalk and knowing full well what had happened, scratched his head and exclaimed, "They sure don't grow 'em like they used to, do they Ed?" *—Courtesy Zeke Zanoni.*

Cool It

Winters were the hardest. The tram was knocked out in February of 1938 and iron workers were summoned from Denver to repair it. The county opened the road, such as it was, and kept it open so supplies and materials could be brought to the different tower sites. The Denver men were lodged in the Grand Imperial in town. At that time, the owner spent little money heating the downstairs and even less heating the men's rooms upstairs. One morning the boss gathered the workers around him in the lobby: "Boys, we lost one of our men last night. Froze to death." Harry, on the outside of circle and not likely to miss a beat, spoke right out: "I'm not surprised. What room was he in?" The tram was back up and running by May. ❋

This is the catwalk beside the tram house that ran up to Main Level. Bob Caine is walking down it on a nice day. Clyde Pitts didn't have such an easy walk here. *Zeke Zanoni Collection*

In February of 1938, the second tower out was struck by a slide and four more towers were also pulled over. Rich Perino Photo

Buried Alive

But in the winter of 1942 it was a different story for Clyde Pitts. He was coming off shift and instead of staying inside and climbing the ladders down to the tram house on that snowy night, he decided to use the catwalk. He had his hands full carrying six to ten cartons of cigarettes. You could say those cigarettes almost killed him because suddenly, in the dark, a slide engulfed him, leaving him buried but alive and able to move only slightly. He fished his knife from his overalls and cut the sleeves from his sweatshirt. After twelve hours of digging, using the cuffs for mittens and the knife for a shovel, he surfaced into the cold first light of morning. The hole he emerged from looked like a corkscrew as he twisted and turned his 6° foot frame to freedom. With little ado, with wrists scratched from where he had cut his sleeves off, he climbed right on the tram and went on down into town and into the bar. He told whoever would listen that he figured he'd had it. Just to make sure he was all right, he stayed in town for almost a week. Then it was back to work. We wish we knew if he quit smoking. ❋

Tram Facts

The Shenandoah tram was used to haul the ore from the mine because the rugged country made year round trucking impractical.

The tram operated from 1930 to 1953 and from 1959 to 1961.

There were fifty or more aerial trams built in San Juan County. The Shenandoah was the only one built with steel towers.

It ran uphill almost two miles, rising 1400 feet. There were twelve towers.

Round bottom buckets carried one ton. Angled buckets carried 1600 pounds.

There were up to fifty two buckets in use at one time. An average of 590-600 buckets of ore were sent to the mill in an eight-hour shift. ❋

John Arietta, tram boss (with the hat), sits with Earl Babcock. Earl was base level trammer who also happened to have been one of the greatest pitchers of baseball in the State of Colorado. Earl gave Bill Rhoades a wonderful old railroad watch just before he died. Zeke Zanoni Collection

A slide splitter was built after a snowslide knocked the tram out in 1938. (The word avalanche was hardly ever used back then. That term was reserved for something huge—that buried a whole town, like in the Andes or the Alps.) *Zeke Zanoni Collection*

The Slide Splitter

Joe Todeschi was born in 1915 at the old Dalla boarding house just across Cement Creek. By the time he was fifteen he was herding horses and mules from town up to the Shenandoah-Dives—by himself. He'd have as many as nine animals in a string. Groceries, coal, kerosene and lumber would make up part of his loads. For that he earned a dollar a day.

But now the year was 1938 and the Sunnyside Mine up by Eureka was closing soon. By June, Joe was out of work. His neighbor across the street was Algot Andrean, tram superintendent, and so, determined to get a job, Joe knocked on the door. Algot was shaving and his wife answered the door. Hearing the voices, Algot joined them and barked, "What do you want, kid? We ain't putting anyone on." Joe needed the job and persisted until Algot relented. The next day Joe was headed up the hill to join his step-dad, Tony Bass, along with Eddy Valentine, Sam Manick and the mason, Carlo Palone, as they constructed the new slide splitter. Joe was hard at it, working closely with the talented mason. The weeks were passing and the slide splitter was growing. Once Charlie Chase came by on one of his frequent mine trips and inspected the growing piece of work. "Don't smooth the rocks on the inside, boys. It will all be covered over. You don't have to make it so nice." Carlo understood English even though he didn't speak it. He picked up his tools and stomped off to the tram house. In his mind the only way to do a mason's job was to do it right. Charlie quickly turned to Joe, "You go tell him to come back. He's doing a beautiful job. Just do it the way he wants. Finish the job." Carlo did come back and the job was finished when the cement cap, using 137 bags of cement that had come up on the tram, was poured. The completed slide splitter stands today, as well as the tower it was built to protect—a testimony to the quality of the work done by the men who built it in 1938. ❋

The slide splitter, after uncounted avalanches, still protects that tower 56 years later. *John Marshall Photo*

"It was probably April, 1950. Bob Caine and I were coming down off shift. When we hit the bottom they told us to call the mine. Turned out

THANKS

they wanted to change the bearings in the compressor and change the oil. It had started running hot. We went back up. That compressor was the heart beat of the mine. When we got done, Vern Hamish comes out and says, 'C'mon over and eat supper, on me. I'd like you to stick around and make sure she's running okay.' Well, after dinner the compressor was doing fine. Bob and I were ready to go, but the tram was shut down. 'When we get ore, you'll get a ride,' Benny Tomassi told us. Turns out Vern Hamish wanted a ride too, and being the superintendent, he got what he wanted. Vern calls the mill. 'Get this tram running. I'm coming down with Bob Caine and Joe Todeschi.' The buckets started and we got on. It was after eight on an April night and a cold rain was falling, threatening to turn to snow at any moment. When we got down to about where the old Iowa Mill had been, that cable started swinging—way up and then wa-a-y down. We thought we were goners but finally the cable stopped and things began settling down. Too much so. Because there we sat until 1:30 a.m. We were cold, I mean, jeeze, we were cold. At last we could see lights coming up the canyon. It was Joe Arietta, the tram boss. He was checking the cable and towers. He'd stop every little bit and shine his light.

"I may be confusing you guys, but hell, I'm confused myself. Everything I told you, as far as I know, is on the up and up." Joe Todeschi easily recalls events from 50-60 years ago in the comfort of his kitchen, 1994. *John Marshall Photo*

'Anyone on there?'

Finally, he gets near us.

'Damn right. About sixty feet above you,' we answered.

'If I can get you a rope, can you guys slide down?'

'Hell no, we're froze!'

'Well, what do you want me to do?'

'Run us into that next station. We'll get out on our own.'

So that's what we did. We climbed down and Joe had somebody else with him. That gentleman was carrying a quart of whiskey. We did some justice to that and Joe says, 'You guys ready to go home?'

'Well, yeah, but we got another man behind us.'

'Who, who?'

'Why Vern Hamish, the Boss, that's who!'

'Oh my God! If I'd known that, I could've been up here two hours sooner!'

The darkness covered the looks on my face and Bob's, but so help me I'll always remember that."

In 1969, Joe Todeschi was working again with Joe Arietta, this time in Maggie Gulch, when the cat Arietta was driving backed over a 100 foot cliff and landed on top of Joe Arietta, killing him.

Riding the tram on a sunny day was one thing—on a stormy night it could be entirely different. Paul Bingel goes up with a fellow worker, 1945. *Wilma Bingel Collection*

IMAGINE THE MINERS WHO RODE THESE TRAMS TO GET TO WORK. Day in and day out. In the dead of winter. In the nighttime. As tough a commute as you can find.

THE TRAM FROM THE MINE LAST RAN IN 1961. The Mayflower Mill continued on until 1994. A jeep or hiking boots will give you these views, but the buckets hang empty, swaying in the wind between the few remaining towers still standing. The Shenandoah-Dives Mine and its companions in Arrastra are closed. Z.M.K. briefly explored the mine in 1989, but one man was killed and shortly after that they pulled out. In 1992, Standard Metals blasted the main level closed—Shenandoah-Dives is no more. The Silver Lake District in the basin above is quietly assimilating the bones of the once busy mines. Gone are the shouts of the miners, the sounds of the donkey trains, the hum of the trams, and the beats of the engines. It's the marmot's whistle that greets the sunrise in this landscape booming and bustling with nature now instead of mining. ❈

Stuck in a bucket over the Animas River, 1961.
Andy Hanahan Photo

Andy Hanahan in the Fourth of July parade, 1994. ***John Marshall Photo***

Hanging Out

George Bingel was the mayor and also ran the tram in that last year of operation—1961. The mayor's job never did pay anything and George loved machines. Andy Hanahan tells this one. "Me and John Cook were wanting to go to Silver Lake the next day and we asked George if we might ride the tram. See, that would put us right where the trail to the lake starts up. 'Sure, boys. You sign these cards and we'll get you up.' Next day, we showed up and climbed aboard. Trouble was, we'd started late and by the time we'd gotten to the trailhead a few avalanches had rumbled down and the weather wasn't so hot. We could see a lot of our walk was going to be in the snow. Our enthusiasm, unlike the snow, was melting fast. Just then they brought a fellow out of the mine with a twisted, broken leg. They loaded him in a bucket, splinted leg and all. One of the men who'd seen us standing around said, 'Here's your chance boys, if you want to get on down. There won't be no ore coming down until late afternoon.' We didn't waste much time thinking about that and jumped in three buckets behind the injured miner. When we got down to where we were hanging above the Animas River, the tram stopped. I turned to John, 'Must be they're unloading that poor guy. We'll be going in a minute.' Ten minutes later I wasn't so sure. A half hour later we were wondering if we jumped would the water below us be deep enough. Well, George knew there wasn't any ore coming down until later on, and still thought we were on our way to Silver Lake. Fact is, nobody knew we were hanging out there, on the highest part of the tram, dangling above the river on a cold, wet spring day. Almost four hours later the tram finally started up. George knew about twenty-five buckets would go by before the first ore would be coming in. Of course, we were only three buckets out and pretty quick we spotted George kind of slouched down over there in the tram house. 'Hello, George,' we shouted in unison. Man, he jumped six different directions at once. The next time we went to Silver Lake, I know we walked." ❈

The mill itself ran on ore brought down Cement Creek from the Sunnyside Mine of Standard Metals Corporation until 1991. *John Marshall Photo*

The Shenandoah-Dives Mine, a large employer, closes. Quiet and depression appear. Then the Mayflower Mill, shut down for a while, begins to run again on ore from a new and even larger employer—Standard Metals.

The tram house in 1994. *John Marshall Photo*

Left to right: Zeke Zanoni, Rich Perino, Vince Tookey, Joe Todeschi, Jim Cole, Bill Rhoades and Gene Miller. These are some of the men involved in mining most or all of their lives, without whose input and help this book would not have been possible. *John Marshall Photo*

The view from a tram bucket heading down into the mill located just off the road to Howardsville, 1946.

Rich Perino Photo

THE MAYFLOWER MILL

All the upper workings were created for one purpose—to get ore to the mill, in this case, the Mayflower Mill, built in 1929 with the rest of the Shenandoah-Dives Company. The mill served the Shenandoah-Dives well from 1930 to 1953. The Shenandoah mined a lot of rock, not ore, but thanks to the steady hand of Charles A. Chase the mine and the mill ran for more than twenty years. The tram carried the ore over the Animas River passing high above the train tracks. The processed ore was trucked to the railroad head in town two miles away and later to smelters around the country. *Mining World* reported

Forty-eight years later, the parts are there but the motion has stopped, 1994. *John Marshall Photo*

The Mayflower Mill Complex with apartments, offices and tram house off to the far right. *Photo by Sanborn, Tom Savich Collection*

on the operation of the mill as they saw it in 1942:

"The ore of the Shenandoah properties is a gold-silver-lead-zinc-copper complex, and selective flotation in the 700 ton mill produces gold, silver, lead and copper concentrates for shipment to the Arkansas Valley smelter of the American Smelting and Refining Co. at Leadville. A zinc product, said to be the highest grade in Colorado, is shipped to the smelter at Amarillo, Texas. A gravity concentrate containing iron and gold is also shipped to the smelter at Leadville. The Shenandoah-Dives Co. also buys custom ore from smaller Silverton producers, but tonnage from this source has dropped off materially in the past two years owing to decreased activity of the smaller operators of the district. Arthur J. Yahn is mill superintendent, with Dan M. Kentro as assistant superintendent and metallurgist."

There was a lot of activity at the mill besides crushing ore and extracting minerals by flotation such as assaying during the milling process and fabricating equipment in the machine shop. We also find the mine administrative offices and more at this site.

Sharon MacFadden and Fenrick Sutherland sit at their desks in their office by the mill in 1940. It was in the building next to the tram house. The top floor was offices; the second floor was the assay office and the bottom floor was apartments for the mine superintendent and the watch man. Sharon was a general accountant and Fenrick mostly did payroll. Charlie Chase had his office in this building as well.

Vince Tookey Collection

Mayflower Mill of the Shenandoah-Dives Mining Company San Juan County, Colorado

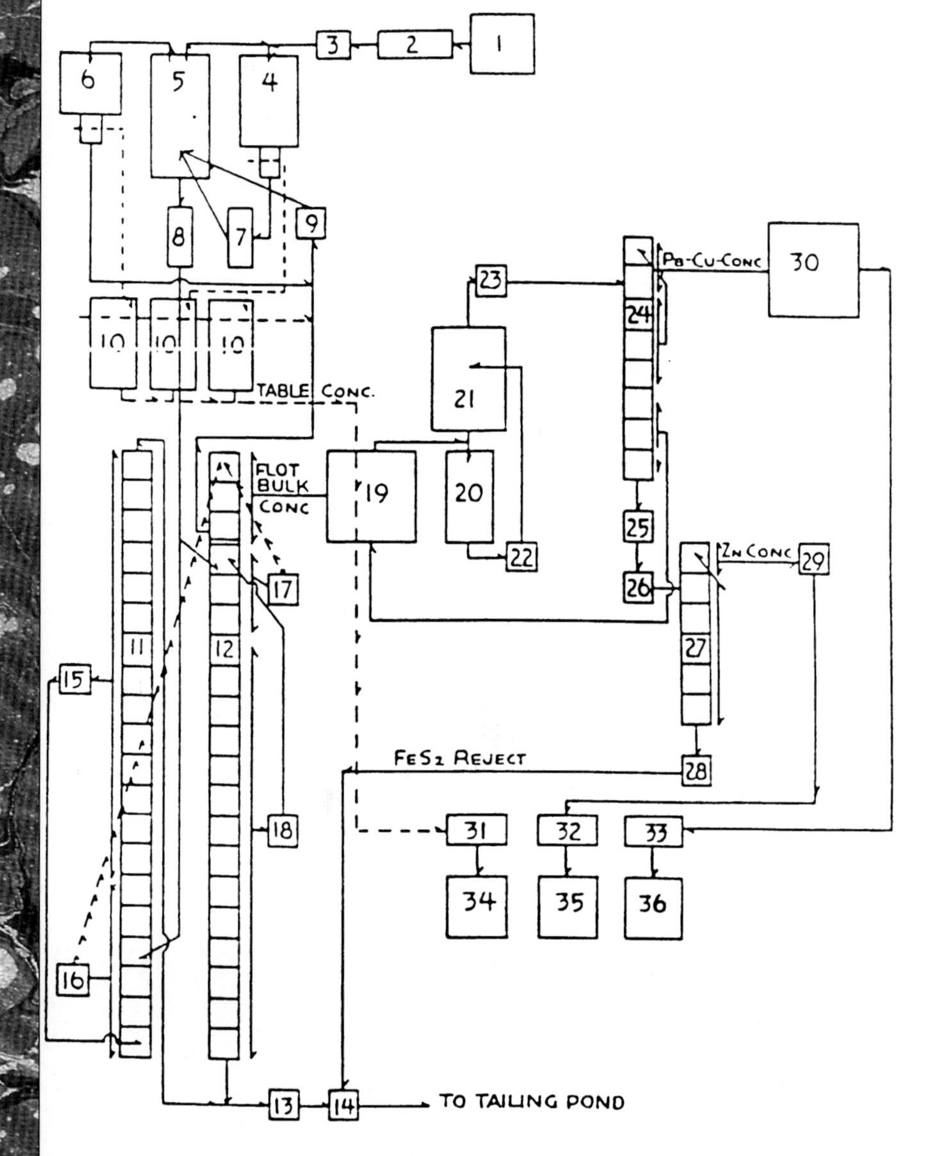

1. 1,200-ton fine ore bin
2. Pan conveyor, 24"
3. Symons short head cone crusher
4. No. 86 Marcy grate ball mill
5. Dorr quadruplex classifier 12x26'
6. No. 64 Stearns-Roger ball mill
7. Bucket elevator 35"x22"
8. Trash trommel, 9-mesh, 2°x6'
9. Belt elevator 24"
10. Three No. 6 Wilfley tables
11. 20-cell No. 21 M.S. flotation
12. 20-cell No. 21 M.S. flotation
13. Liberty Bell type sampler
14. Hydroseal pump, "B" frame size
15. Wilfley pump, 2"
16. Wilfley pump, 2"
17. Wilfley pump, 2"
18. Wilfley pump, 2"
19. Dorr thickener, 35'x10'
20. Stearns-Roger ball mill, 4'x10'
21. Esperanza-type classifier, 6'x16'
22. Wilfley pump, 3"
23. Wilfley pump, 3"
24. 8-cell No. 18 Denver flotation
25. Denver conditioner, 3°x5'
26. Wilfley pump, 2"
27. 6-cell No. 18 Denver flotation
28. Liberty Bell type sampler
29. Denver 1" concentrate pump
30. Dorr thickener, 35'x10'
31. Settling box, 3'x6'x2'
32. Dorr filter, 2°x6'
33. Dorr filter, 5'x10'
34. Table concentrate bin
35. Zn concentrate bin
36. Pb-Cu concentrate bin

Mining World (1942)

Interestingly enough, in the late sixties Oliver Hower realized that by moving the Symons crusher (#3 in the diagram) over to the crushing plant, the capacity of the mill would go from 600 tons a day to almost 750 tons a day. It was Aldo Bonavida and his crew that were able to implement the idea and realize the increased capacity.

THE TECHNICAL LAYOUT OF THE MILL'S OPERATION is detailed in the diagram above for those familiar with the milling process. Put more simply: the mill takes the mined rock, crushes it to a sand-like consistency and through various chemical precipitation processes, floats off and recovers any valuable minerals known to be in the ore. Minerals such as zinc, lead, copper or silver. But the most important operation is the recovery of the gold—soon to be described in the amalgamation process. This flow was changed somewhat in later years when the Sunnyside's ore began to show up.

In the picture at right are the float cells in the Mayflower Mill in 1941. The lead circuit was on the left and the zinc circuit on the right. The concentrates, in these early years of the thirties went by truck to the narrow gauge railroad in town. From there they went to Alamosa and north or south on broad gauge rails. In the seventies the float cell area was more than doubled. Vince Tookey Photo

ASSAYING IS DONE AT SEVERAL STEPS IN THE PROCESS as *Mining World* reports in 1942:

"A completely equipped assay office and laboratory is manned by two graduate engineers, Gayle Farnham and James Cole, with two men preparing samples. About 100 fire assays and 130 to 140 wet determinations are made each day. Laboratory equipment includes a sample drier, crusher, Davis coffee mill, and McCool pulverizer, together with a DFC oil-fired, two muffle-furnace."

A longtime mill man and immigrant from the "Old Country" was nabbed collecting gold concentrate in a Prince Albert can as it came off of a Wilfley Concentrate table in the Mayflower Mill. The supervisor immediately reported the incident to the General Manager, Charles Chase. Mr. Chase, being a compassionate man who hated to fire anyone, transferred the employee to work at the tailings pond instead.

After working the pond for several weeks, the employee went back to the mill office after shift and asked for the General Manager. Upon entering Mr. Chase's office and catching him at his desk, the man extended his open hand which contained a good portion of gold dust. With his very broken English he said, "You see, Senior Chase, you loosa the high grade in more place-a than the shakin' table!"

Unfortunately, Charles Chase's reaction is unrecorded, but one might assume his worker was soon out prospecting for a new job. ⁂

—*Zeke Zanoni*

Jim Cole's father, Jim Cole, Sr., works in the Mayflower's assay office in the forties. He's doing wet analysis on hot plates under the hood, 1944. Jim used to ride his bike out to see his dad and maybe split a lunch. As Jim grew older, he found himself doing assay work too. Jim Cole Collection

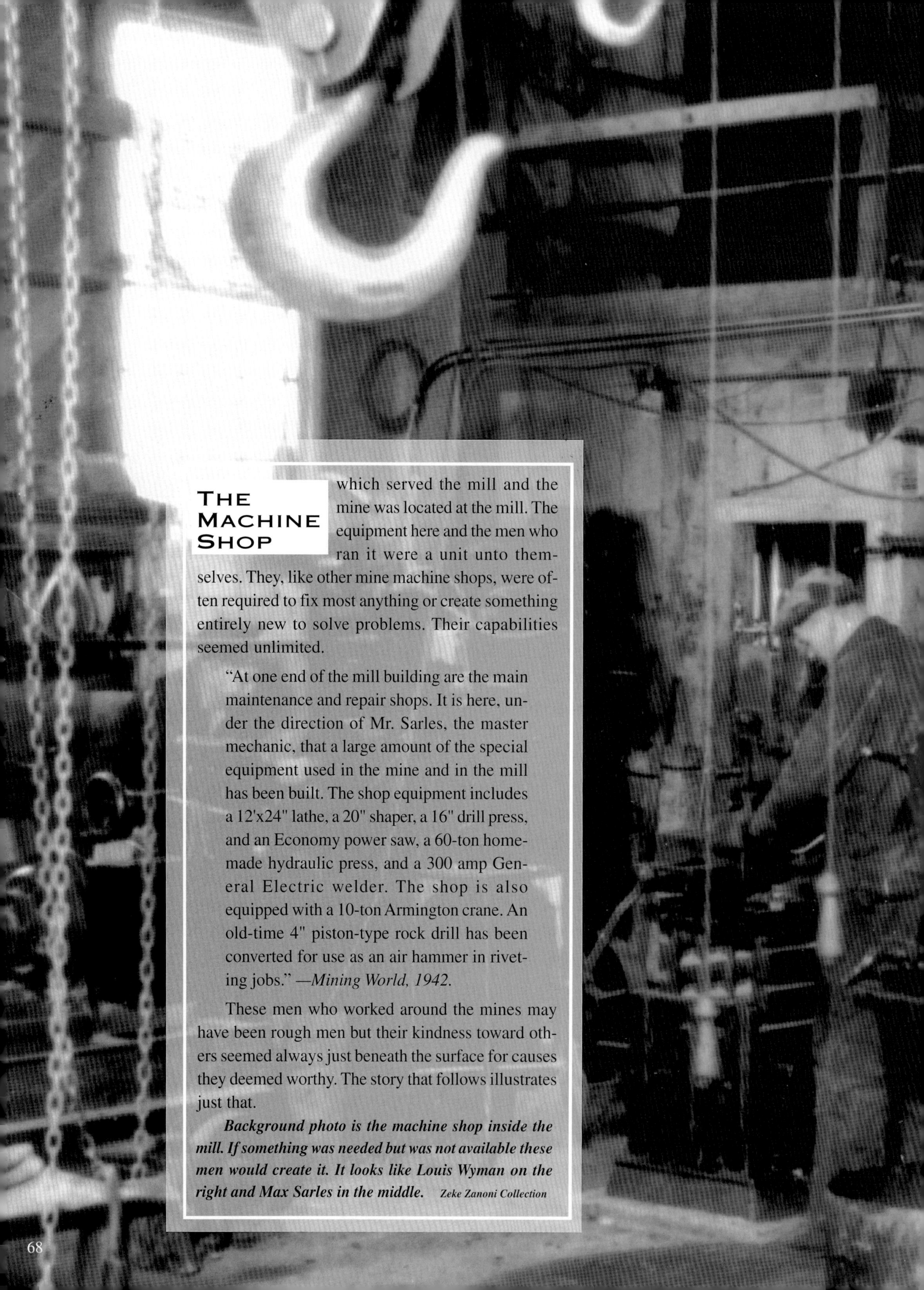

The Machine Shop

which served the mill and the mine was located at the mill. The equipment here and the men who ran it were a unit unto themselves. They, like other mine machine shops, were often required to fix most anything or create something entirely new to solve problems. Their capabilities seemed unlimited.

> "At one end of the mill building are the main maintenance and repair shops. It is here, under the direction of Mr. Sarles, the master mechanic, that a large amount of the special equipment used in the mine and in the mill has been built. The shop equipment includes a 12'x24" lathe, a 20" shaper, a 16" drill press, and an Economy power saw, a 60-ton homemade hydraulic press, and a 300 amp General Electric welder. The shop is also equipped with a 10-ton Armington crane. An old-time 4" piston-type rock drill has been converted for use as an air hammer in riveting jobs." —*Mining World, 1942.*

These men who worked around the mines may have been rough men but their kindness toward others seemed always just beneath the surface for causes they deemed worthy. The story that follows illustrates just that.

Background photo is the machine shop inside the mill. If something was needed but was not available these men would create it. It looks like Louis Wyman on the right and Max Sarles in the middle. *Zeke Zanoni Collection*

—Another story from Louis Wyman's Snowflakes and Quartz

The Old Man

Every resident mine manager is known as "the old man." Age has nothing to do with it. It's a title bestowed by the crew, whether they like him or not. It always carried a note of respect for his authority. His signature on the paycheck backs it up.

In the early days of the Shenandoah Mine, the machine shop crew at the mill became used to the old man coming and going through the shop. It was the shortest way between the mill and general offices. Often he'd stop and comment on whatever work was in progress. There was very little that went on about the plant and mine he wasn't well informed on.

Most of our side activities he chose to ignore, but when he found something he couldn't tolerate he'd come down hard on us. He was recognized throughout the Southwest as one of the best managers in the business. And on matters of policy most everyone figured he was reasonably fair.

The shop crew at the mill developed a little side line, a kind of good neighbor thing, that just seemed to have sprung up by itself. There was no money in it. The miners on their way to work in the morning would leave some household gadget in the shop to be mended. Then they'd pick it up after shift on their way home.

Quite often when we came to work we'd find a youngster's toy tucked away behind a tool box or under a work bench. Bicycles, tricycles, little red wagons, and even a little girl's doll buggy came our way. With each toy came a silent plea—"Please, mister, will you fix it for me?"

We had to keep the stuff out of sight as much as possible, even though we suspected the old man knew all about our side line. We watched his comings and goings, and when he went to the mine on inspection trips, we'd pull out our backlog of toys and try to get caught up with our work.

One day a new angle developed. I glanced up from my work and saw a woman standing in the doorway. She seemed unsure as to whether she should come in or not, so I went over to see if we could help her in some way. She said her husband worked at the mine. They thought if she brought their little boy's leg brace in, someone might be able to fix it for him. She handed me one of those monstrosities of bent iron and leather

This is M.R. Sarles, master mechanic. Sarles and his workers could, and did, create many wonderful mechanical works of art. This is a motor he made at the mill to run the locomotives in the mine. *Zeke Zanoni Collection*

straps which doctors and technicians use to harness kids crippled from polio.

The shop crew got to know that brace and its owner, Tommy, quite well. He was a rough one, twelve years old, and he'd try anything. Though his leg must have given him real pain at times, I never heard him whimper. Bicycles, basketball, baseball or anything else where the action was fast, was just his meat.

Zeke Zanoni Collection

Sometimes we almost gave up trying to keep him on two legs. It got so I thought leather and iron weren't strong enough to hold him. To Tommy, the world was a great big wonderful place to have fun in. And a leg brace—no handicap at all.

On one occasion he came limping into the shop, climbed up on the work bench, and informed me, "I broke it."

"So, what's new about that?"

"I fell off a burro. He was running fast. But I didn't hurt my other leg."

"Well, in that case we should be able to get you back on two legs again. What are you planning to do? Ride in the Labor Day Parade?"

It happened that the old man was on a trip to the mine and would be gone for a time, so we didn't have him to worry about. The shop welder mended the broken part for us. Then Tommy and I clamped it in the bench vise while we finished it off and fitted a new pin to the hinge joint.

For convenience sake, I kept Tommy perched up on a box on the work bench, so I could find him when the job was finished. He always gave me a run-down on important events and could ask questions about everything he saw in the shop and mill faster than I could answer them.

Suddenly he fell silent. I turned to see what had attracted his attention. The old man stood back of me, taking in the whole performance.

"Well," I thought, "Here we go, caught at the scene of the crime."

"Are you repairing the lad's leg brace?"

I told him I was and introduced Tommy. The old man shook hands with him and addressed him with the full title of "Mister." Tommy sat up straight on his box and grew an inch or two.

"Louis," the old man said, "Spare no effort. Do the very best job on that brace you can." Then he shook hands with Tommy, said good-bye, and went back to his office.

Yes, the old man was boss. He hired and fired, he carried the weight of responsibility for a big outfit on his shoulders. But he could still find time to be concerned about a little boy with his leg in a brace. ❋

THE MAYFLOWER MILL CLOSED IN 1953 AFTER THE 44 RAISE COLLAPSED and the rest of the Shenandoah-Dives mine proved unable to come up with new ore. After sitting idle for six years Standard Uranium fortuitously showed up and reopened the Shenandoah-Dives returning to the Silver Lake section of the mine, hoping to prove new reserves of ore. This reopening was a two-pronged operation. The bulk of their energy was focused up Cement Creek where they built the American Tunnel—driving under the workings of the closed Sunnyside Mine. The Sunnyside had operated until 1938 out of Eureka as Colorado's largest gold mine. This new tunnel enabled the Mayflower Mill to reopen and a new era of mining began in Silverton that lasted into the eighties, and even the early nineties. The first ore from the American Tunnel was hauled by truck down Cement Creek to the mill in 1961. For the rest of the sixties, 600 tons a day came from the Standard Metals project to be processed at the mill, however little profit was realized. Now called Standard Metals, the Sunnyside Mine and Mayflower Mill struggled into the seventies and, after expanding in 1975, the mill processed 1,000 tons per day, six days a week. Once again, the Sunnyside had become the richest gold mine in the state. ❋

Tom Savich left the San Juan County Road Crew to go to work at the Mayflower Mill in 1973. He did all kinds of mill work, ran the floats, did repair, and drove heavy equipment. His father, Pete, worked up on the hill at the Shenandoah in the thirties. If you are ever fortunate enough to be invited for a tour of the museum he's created beside his home, don't turn it down, 1994. *John Marshall Photo*

amal·gam·ation 1. the action or process of amalgamating, uniting. **2.** the dissolving of a metal in mercury to form an alloy.

Webster's version needs elaboration. Turning ore into bars of gold was the last step of the many difficult and laborious mining operations. Pat Donnelly was willing to put her ten-plus years of experience as an amalgamator into words to help us understand this complicated process.

From Rock into Gold

"I was trained to be an amalgamator by Fred Andersen who had worked at the mill for many years. He was a good teacher and friend. Yvonne Conrad also provided part of my training. However, amalgamators are somewhat like alchemists of old; they provide you with the information you need for the basics but keep the rest to themselves. Therefore you must spend a great deal of time reading and researching. The amalgamators' secrets, it seems, went with them to their graves."

The amalgamation area was a large room with two floors, the upper floor was made of grating and the lower one of concrete with two inches of acrylic, and had the California plates. The amalgamation barrel, the retort, and the reagents were stored on the upper floor which was not heated because mercury was kept there.

There were several steps processing and extracting the gold from the ore. This particular ore came from the Sunnyside up by Gladstone. First, it went into the mill in the coarse ore bin where it was fed into a jaw crusher to break it into smaller rocks. From there it went to the fine ore bin where it was fed into the rod mill which ground it to a fine, sand-like consistency. From there it was fed to the ball mill for an even finer grind. As it came out of the ball mill, it flowed across the jigs. The constant back and forth motion of the jigs at a rapid rate would sift the finer particles of concentrate to a pump below that sent them to a holding tank. The next step used a shaker table, a specially built table, eleven feet long and seven feet wide, with a wheel to tilt it at an angle. The top was made of chemically treated rubber with baffles to slow the flow of concentrate.

Herman Todeschi on left and Fred Andersen at Herman's retirement party in October, 1989. Steve Smith Photo

The concentrate in the holding tank was then fed to the table and with water flowing and the table shaking, the material was separated. The lighter material with little or no value was washed away and channeled back into the circuit. The heavier material such as gold, lead, and silver was channeled into a jig box. A jig box was two feet square and when filled could weigh 900-1200 pounds. The material was very fine and it could take eight to sixteen hours to fill one. Then the boxes were weighed and logged: box number, date, and weight.

The jig box would be taken into the amalgamation room where it was washed into the amalgamation barrel. This is a three foot by five foot ball mill. Reagents would be added to clean the gold so it would be ready to receive the mercury. After grinding the concentrate with the reagents for several hours, about eighty pounds of mercury would be added. The more gold reported in the assay, the more mercury was added. This is where some wizardry and experience was definitely needed. The new chemical combination was ground for several more hours at a slower speed.

The next step is called elutriation: 'to purify, separate, or remove by washing.' By going to the lower level, one prepares to recover the mercury. If everything has been done right to this point, the gold is combined with the mercury. Under the amalgamation barrel is a box with a cone on the bottom. A glass tube, two inches in diameter and about twelve inches long is attached to the cone. At the end is a ball valve with a water line placed just above it. A five gallon bucket in a tub is placed under this and the water is turned on. The amalgamation barrel is unplugged and then begins to rotate slowly. As the barrel turns a small amount of the liquid inside comes out and goes down the cone into the glass tube. The ball valve is opened just enough to let he mercury out. If the water pressure is correct, the mercury, with clean water, will flow into the bucket and the other reagents and unwanted material will flow over the California plates located below. The plates were positioned to catch any gold/mercury that might escape during elutriation. The overflow matter is sent on to the lead thickener tank.

When the barrel has emptied and been turned off, the bucket with the mercury is taken to an air press where air is added. This is where you really start to become the center of attention for the profits of the mill are dependent upon your actions here and the amount of gold that is realized. Wait about twenty minutes and open the press. The mercury will be in a solid form because it is saturated with gold. This product, called amalgam, is weighed, recorded, put in a container and locked in the safe. When enough material has accumulated it's time to retort.

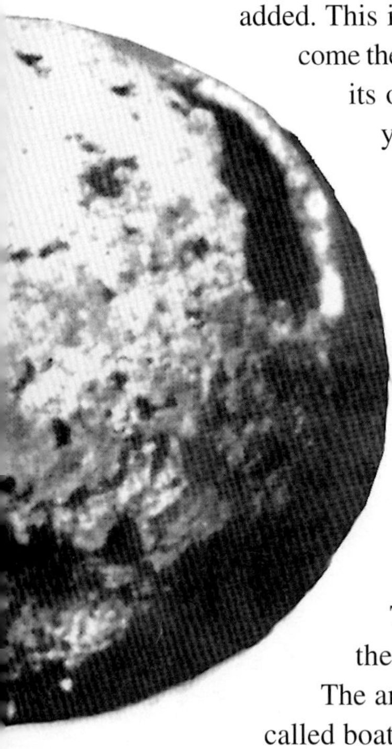

The mercury is driven off from the gold from the retort process. The amalgam is placed in containers called boats. They are half-moon-shaped, about four inches wide and six inches deep, and are made of °-inch steel plate. Using hydrated lime and water, a paste is made and the inside of the boats are coated and allowed to dry. Then the amalgam is placed in the boat and pounded down until the boat is full. One to six boats can be retorted at one time. The boats are loaded into the retort, the door is lined with fire clay and secured with locks. The boats are cooked for twelve hours at 1200 degrees Fahrenheit. A fine pie indeed. Then the retort is turned off and allowed to cool to about 200-300 degrees Fahrenheit. What is left is a gold sponge.

The weight of each boat is recorded then three sample holes are drilled and sent to the assay lab. Now the sponges are packaged and numbered for shipment. They are then sent to a refinery in California where the gold is refined to 99.9% purity.

While performing all of these duties the workers were under constant camera surveillance. There were cameras on the shaker table, the safes, the retort, the lower level, and the doors. These filmings were monitored by security guards. During shipping a security guard was present at all times. Shipping was done in small black cans that had a bail and were wrapped with wire that was welded. The cans were crimped around the edges like a bucket of roofing tar. The post office in Silverton sent them out by registered mail, sometimes five cans at a time and sometimes insured for a quarter of a million dollars. But even with all the precautions taken, the mill was the scene of three major thefts: one in June of 1978 for about $20,000 dollars, in July of 1981 for $120,000, and one a year later in 1982 for $110,000. The perpetrators were never caught although the FBI, CBI and the local sheriff did their collective best. ❋

—Contributed by Pat Donnelly

(center) A gold sponge is bowl shaped and could be five inches or six inches thick and weigh up to thirty-five pounds. It was eighty-five to ninety-five percent pure gold. If the ore being worked ran 0.25 troy ounce gold per ton, then four tons of milled rock would make one troy ounce of gold. Precious metals, especially gold, are measured in troy weights: twelve troy ounces equal one pound. In our example, at 0.25 ounce per ton, it would take forty-eight tons of rock to make one golden pound. So, one thirty-pound gold sponge would contain 360 oz. of troy gold and would have come from 1,440 tons of rock. And if gold were $400 dollars per ounce at that time, that thirty-pound sponge would be worth about $144,000 dollars. The Sunnyside Mine in the seventies and eighties milled 1,000 tons of rock in one twenty-four hour period. Remember 0.25 gold per ounce is good ore. Some ran higher, most ran much lower. And remember, these figures only hold true when the gold is actually in the rock you're mining at that time!

Allan Bird Photo, Silverton Gold, Fig. 138.

Pat Donnelly (1994) shares her knowledge gained by doing—from the shovel to the rod mill and on to amalgamation, "I had ten years in the mill by 1991. All in all, they were an experience I wouldn't have missed for anything. I made a lot of friends and I miss them very much." ***John Marshall Photo***

Eric Olin, the mill superintendent and the first person on the left, took this picture in 1984.

KNEELING LEFT TO RIGHT:
Eric Olin, Mill Superintendent
Gloria Sandell
Fred Manzanares
Susan Langley
Tom Howarth
Dennette Gonzales
Evan Buchanan
Dave Wiggins
Barney Darnton
Chip Wagner
Tom Savich
Allen Brown
Herman Todeschi
Steve Smith
Don Norris

STANDING FIRST ROW:
Jack Lowery
Danny Chacon
Jack Wyman
Gary Ficklin
Bill King
Jerry Sandell
Joe Todeschi
Pat Donnelly
Everett Gardner
John Zarkis
Greg Leithauser
Jeff Salazar
Roy Young
Bobby Robbins
Mike Manzanres
Bill Dotter

STANDING SECOND ROW:
Richard Ostendorf
Steve Gray
Chris Kennedy
Randy Snyder
Jean Willis
Max Slade
Dave Ferguson
Dave Dillon
Mark Will
Art Will
Ed Greene
Ken Johnson
Sam Blann
Ralph Rae
Jamie Gurule

LAST FOUR ON LEFT AT BACK:
Jim Kaiser
Doug Murray
Dennis Norton
Ed Brown

IN THE SUMMER OF 1981 JACK LOWERY AND SAM BLANN WERE WORKING ON DAY SHIFT IN THE CRUSHING PLANT OF THE MILL.

LAID OFF FOREVER

Steve Smith was the shifter and Barbara Morris was the safety officer, now busy on an inspection of the mill. Suddenly, Jack's voice came over the squawk box. "We need help in the crushing plant. Help in the crushing plant. Right now!"

Steve ran to the plant at full speed with Barbara close behind. They arrived finding Jack bent over a prostrate Sam Blann. Jack had already killed the power so the plant was pretty quiet. Unfortunately Sam had done what was a fairly common practice in the mill. The conveyor belts were slipping because the rock they were crushing was soaking wet. Sam had taken a stick with a rag wrapped around it like a torch and soaked that cloth in pine tar. Then he'd reached past a guard to swab that stick across the head pulley and make the belt stick. With no warning, the pulley had grabbed the stick and given Sam no chance to let go. His arm was pulled in and ripped off, taking his shoulder and shoulder blade right out, too. Tough as Sam was, he had the grit and the presence of mind to walk the length of the belt (about seventy feet) to catch Jack's attention. If he had tried to yell, Jack would never have been able to hear him. If he had passed out Jack would never have been able to see him.

When Steve got there, Jack was stuffing rags into this huge cavity in the man's side, trying to stop the bleeding. It took both Steve and Jack's hands pressing down to slow the flow of blood. The deck was only a couple of feet wide and eight feet off the ground. By now, there were ten or fifteen men around, and in order to get Sam out of there, Steve and Jack had to lay on the stretcher with him, pressing on the cavity the whole time. They went to the ambulance which had come from town, all three lying on the stretcher. At this point, Greg Leithauser relieved Jack who went to get Sam's wife. Someone else had gone down under the conveyor belt to get Sam's arm and put it into the ambulance in a trash bag. Allen Nossaman was the driver. Janet Sharpe and Marge Baudino were the E.M.T.s. They all shot into town where Tom Galbraith from the French Bakery met them in the middle of Greene Street with a trash can full of ice which was promptly poured into the bag holding the arm.

Then began an extremely fast twenty-five minute ride to Ouray where the ambulance was met in the ball park by a helicopter. Sam had never lost consciousness and was even able to do a little joking as the technicians kept questioning him, trying to keep him from going into shock. Steve and Greg were still using both hands on his shoulder, Steve especially, was now soaked with blood.

And yet, after only two days in critical condition in Saint Mary's Hospital in Grand Junction, Sam was on his way back to life, a life that would be forever changed. His arm never was reattached. The mill which had been shut down that morning was ordered back to work for the swing shift. Six months later, Barney Darnton hired Sam back to prep samples. Less than two years later, he was laid off, forever.

Steve Smith with his son Cory in 1988 at Fuller Lake above Ice Lake. The picture was taken over the ears of Trigger, Marvin Blackmore's horse. That same night Cory blew off the thumb, index and middle fingers of his left hand with a blasting cap on a bottle rocket. Three earlier rockets had been successful...

Steve was no stranger to accidents as the drama above reveals. Sam Blann might have wished it was just a thumb. *Steve Smith Collection*

The Mill Song by Dan Bender

When I was young and out of work I heard about this mill,
So I rustled 'round, got hired on and moved on up the hill.
I told my Boss on my first day
of mills I did not know.
He smiled wide, said, 'It's O.K.'
We'll teach you the banjo.

Big Bubbles
No troubles
Watch the metals flow
First we break it
Then we shake it
In some air we blow
Makin' milkshakes
out of mountains
Just to get out all the gold
Workin' in this gol-dang mill
is gettin' pretty old

A year went by, I shovelled hard and seldom was I flustered
'Til the Boss claimed he promoted me and put me on the crusher
Rail and timber don't crush too well
that fact I learned real fast
All year I dreamed of other jobs
Each shift I wished my last.

Chorus

Rods and balls and crater x and densities of muck
Now that I run the grind circuit, I feel a change of luck
Sweat and freeze, oh what a breeze this mill work's got to be
I thought that I was set for life, the bad times were not seen.

The markets fell, the layoffs came, I hadn't planned for that.
So I hang out, I drink lots of beer and I'm getting soft and fat.
They started with new management and swore they would not fail
but I didn't know that when they called me back, they'd put me on the tails.

Big Bubbles
No troubles
Watch the metals flow
First we break it
Then we shake it
In some air we blow
Makin' milkshakes
out of mountains
Just to get out all the gold
Workin' in this gol-dang mill
is gettin' pretty old

From the CD **Friends in High Places,** *Dan Bender and Ray Liljegren* **copyright © 1994.**

"I was the only working production miner living in the county, driving 130 miles round trip every day. Then six months ago another guy living here got a job at the same mine. In one day we doubled the number of working miners living in San Juan County, 1994."
Dan Bender at home, 1994. *John Marshall Photo*

Bill Melcher levels the ground near the Mayflower Mill preparing a highway pullout for interested travellers, 1994. The plaque below is part of the display.

The mines of Arrastra Gulch lay behind him, ahead of him and beside him as he levels the area around him.

John Marshall Photos

"I feel like I'm building my own tombstone." And in a way, maybe he was—for all the hard rock miners—and yet, maybe not…

Mayflower Mill Tailings Ponds

What are Tailings?

In the Silverton region, ore generally contained only about 5% valuable metals, typically a mixture of lead, zinc, copper, silver, and gold. Most mines sent their ore to a nearby processing plant, known as a mill, which crushed the rock into fine particles and then separated the metal by floating it to the surface in specially designed tanks. On its way through the mill, each ton of ore yielded a small amount of metal and a large amount of wet, sand-like material called tailings.

The Mayflower Mill Tailings Ponds

When the Mayflower Mill began operation just up the road from this site in 1930, the mining industry customarily disposed of its tailings by dumping them into the nearest river. But the Mayflower's superintendent and part owner, Charles A. Chase, took an unusual stand: "Because of my personal repugnance for the lack of consideration of the public interest involved in this practice, I undertook to withhold from the Animas River the tailings of our new project."

Borrowing some techniques and inventing others, Chase in 1935 created a successful, on-land disposal system that impounded wet mill tailings in dam-like structures known as tailings ponds. After the ponds dried out, the tailings were left high above the river in the shape of tall, sandy mounds. As the mining industry became more environmentally conscious, impoundment became a standard tailings disposal technique.

The two mounds directly before you are remnants of Chase's pioneering work of the 1930s and 1940s. Among the earliest engineered tailings ponds in the Rocky Mountain mining districts, they make this locale a significant historic site for both the American environmental movement and the country's mining industry.

The Mayflower Mill, c. 1940. The tailings ponds (center background) were located downhill from the mill to allow gravity disposal of the tailings.

(Charles A. Chase Photo, Courtesy of San Juan County Historical Society)

Tailings Pond No.1 in its early stages, c. 1935. The wooden flume deposited wet tailings from the Mayflower Mill. As the solids settled out, the tailings grew into a sandy mound.

(Charles A. Chase Photo, Courtesy of San Juan County Historical Society)

Tailings Pond No. 1, c. 1940.

(Charles A. Chase Photo, Courtesy of San Juan County Historical Society)

This is looking up Cunningham Gulch probably around 1901. The railroad had reached Howardsville and had been extended up to Eureka by 1896. The Green Mountain spur was added by 1905 and a depot appeared. Track was removed and salvaged by 1938. San Juan County Historical Society

Howardsville and Cunningham Gulch

First the Shenandoah-Dives Mine stopped producing and then the Mayflower Mill shut down. This place is getting pretty quiet. Let's jump back in time, use our train ticket and get back on the train. We'll head northeast up the Animas River valley to Howardsville and ride the train on up Cunningham Gulch to the Green Mountain Mill. We'll have a lot of exploring to do in the busy area of Cunningham Gulch. We'll eventually go to the Highland Mary Mine before we come back down and make our way north again to the next town of Eureka.

Howardsville is located in the Animas valley near the entrance to Cunningham Gulch. Good old Howardsville. The town had a post office by 1874 but was never even platted and never incorporated.

THE RAILROAD EXPANSION TO THE MINES WAS THE CATALYST FOR GROWTH AND CHANGE. It moved the most tonnage of material and probably more people than other forms of transportation. And wherever you travelled, it was never boring.

Mickey Logan, now in Durango, born in Silverton in 1904, remembered this story while reminiscing in his shop.

"My uncle used to live in Red Mountain town. I can remember him telling me about a salesman who had never ridden the train before. My uncle told him, 'If you want to get to Silverton you're going to have to help. We've had an awful lot of trouble with the cars jumping off the tracks. Seeing as you're the only one in here, when you feel the car tilt to the right, you're going to have to jump, and I do mean jump, to the left. Otherwise, this car will probably leave the track. We'll be forward helping the engineer.' My uncle checked up on the salesman from time to time. By the time they got past Chattanooga the new passenger was pretty well worn out. But he had done his job well. The car was still on the tracks."

In 1905, a branch of the railroad from Howardsville was built up to the Old Hundred Mill to the Green Mountain Mill in Cunningham Gulch, for a distance of 1.3 miles.

Herman Dalla could remember riding on that branch. Very clearly, by the way.

"Yep. I rode the train up to the Green Mountain Mill. I think they were hauling empty ore cars to drop off at the mill. I was eight years old. I was headed up to live with my sister in Eureka. She needed someone around to help her out. She was married to Phil Antonelli, a blacksmith at the mine, but she didn't like being alone when he was on shift. I can remember Pete Meyers was the engineer; Frank Brown the brakeman, Ralph Plantz the fireman, and Pat Donergan the conductor. Well, I was afraid I'd miss Eureka and end up in Denver. I kept asking Pat how much farther? He kept telling me 'Quit worrying, kid. I'll let you know when we're there.' It seemed like the train went on forever. Are we almost there yet? Then we were. Eureka, 1920." ❋

Herman Dalla, born in Silverton, April 12, 1912, is shown here in his World War II Army-Air Force uniform. Gerald Swanson Photo

Evan Buchanan, engineer in the nineties—1990's.

John Marshall Photo

To get into Cunningham Gulch you would have to pass this structure, the La Plata County Courthouse. Built in June, 1874, by John Burrow, the building stood until August, 1974. The seat of the county when built, it housed court functions, county sheriffs, and county clerks. But the county seat was moved to Silverton in September of 1874 and the county was renamed San Juan County. Close—we might have been North Durango. Ruth Gregory Collection

Ruth Dawson Pruett Gregory is shown at right in Silverton in the summer of 1914. Ruth was born in 1911. Her mother had graduated from Silverton High School in 1904 and had worked with her grandmother in the boardinghouses around town. Grandmother would cook and Ruth's mom would wait tables. Ruth remembers leaving the Green Mountain Mill around 1916 to go to Silverton. They didn't take the train but rode horses instead, with Ruth covered in flour sacks. It was winter and their goal was to see Ruth's first Santa who they found, naturally, riding on the back of a caboose in Silverton in 1916. Two years later, all of seven years old, in 1918, Ruth started school out in Eureka.

Ruth recalls about the cabin pictured above, "My mother went to school in this building." Note the railroad tracks heading into Cunningham. Ruth Gregory Collection

This is looking southeast from Howardsville up Cunningham Gulch. The Old Hundred Mill in the foreground, from 1904 through 1908, had three separate trams that led to a fascinating collection of mine buildings perched high on the cliffsides above the valley floor. Some of these building are still in existence today. The house on the left was taken out by a mudslide in the thirties. Glenn Sandell and his family, including son Jerry, lived in the house on the right-hand side of the tracks. The next building is the office and boarding house. In 1967, when Dixilyn took over these properties, what was left of all these buildings was removed. Farther up the valley stands the Green Mountain Mill, then the Vertex Mill by the Buffalo Boy tram house at the bottom of Stony Pass and on up the road was the Highland Mary Mine. John B. Marshall Photo

The sure sign of a packer, the man in the lead has his cuff rolled high so he could be easily spotted. The person might just be Louis Wyman, Senior. The building is the staff office and boarding house just across from the Old Hundred Mill. The railroad tracks are barely visible between the pack string and the building, ca. 1906. Andy Hanahan Collection

GALENA MOUNTAIN AND THE OLD HUNDRED MINE

Scott Fetchenhier, Silverton resident and a most knowledgeable gentleman of the mines and geology of the area, contributes the following account of the Old Hundred, the first mining area reached by the Silverton Northern in Cunningham Gulch.

Man's interest in the Old Hundred Mine and Galena Mountain only goes back approximately 120 years, if one doesn't include the occasional Ute Indians on hunting expeditions picking up curious pieces of lead ore for medicine men's healing bags or possibly Spanish renegades prospecting and mining in the San Juans. Such a small block of time becomes insignificant geologically when examining the events that began millions of years ago in this area.

Twenty-five to thirty million years ago, probably for the third time in the history of the San Juan region, intrusion of a large molten rock batholith and subsequent violent volcanic eruptions began to cover the area in flows and ash. As the molten rock chamber was emptied a large block of ground subsided, forming a large crater or caldera. Collapse of the caldera in earth-shattering earthquake activity formed concentric fractures around its rim and large fractures radiating outwards from the rim in all directions. Two of the largest radial fractures to form on Galena Mountain were the #7 vein and the #5 vein. The #7 vein can be seen even from Silverton, running diagonally across the upper face of the mountain as a rusty-colored slash.

Several million years later, rain waters percolating down through the earth encountered the still-cooling batholith. As these waters were heated, they formed chemical compounds that began to leach metals out of the surrounding rock. These hot or hydrothermal waters began to migrate upwards under great temperatures and pressures into the surrounding fractures. As the waters neared the surface, the dissolved minerals and metals began to precipitate out, coating the fractured walls and fragments in between with quartz, calcite, galena, sphalerite, chalcopyrite, pyrite and the precious metals of gold and silver. Subsequent movement along the faults would shatter these deposits which would be cemented later by other ore deposits. These ores were deposited slowly over millions of years. All that remained several million years later after erosion and glaciation, was for the deposits to be discovered by man.

A post card from 1917 shows the mine buildings as they must have appeared during the expansion years of 1904-1908. Postage was a penny. *Tom Savich Collection*

It is hard to say exactly how the two big veins, the #5 and #7, were found on Galena Mountain. The earliest prospectors with the Baker party, in the early 1860's, had found placer gold in the big valley that runs from Silverton to Eureka, but it wasn't until the early 1870's that prospectors began to search for the source of the gold. The earliest discoveries on the veins were made by the Niegold brothers from Germany. Reinhard Niegold may have been in the area by 1872. The veins probably weren't that hard to discover. The talus slopes of Galena Mountain are speckled with quartz float, pieces of the vein that have been broken off from above. The Niegolds probably followed the float uphill until they came across the veins. The veins are also visible to the eye from a distance. In several places in the cliffs above Cunningham Gulch the quartz outcroppings of the vein stick out above the surrounding volcanic rock and can be followed for several thousand feet. The veins contain a fair amount of pyrite, the iron which weathers a rusty brown when exposed to the elements. These reddish stains

The trams ran from the mill to the mine. Here, perched precariously on the cliffs of Galena Mountain were the buildings of the Old Hundred. Among them was a two-story tin bunkhouse, still barely standing today, and the tram house which also contained a blacksmith shop and a work room. The rear of the building, by the workshop, was the entrance to the mine. Operations ceased with the 69th bucket of ore in 1908. Possibly as many as twenty men lived here in 1905, judging by catalogues found in 1953. If you look over your shoulder later when we head up the valley to the Highland Mary Mine, this is what you'd see. The tramhouse is to the right and the boarding house to the left. The curve in the rock above the buildings is the #7 vein, visible from town. *John Marshall Photo*

are painted all over the cliffs of Galena Mountain and would have attracted prospectors as potential targets.

The Niegolds found these obvious outcroppings to be large, shattered veins, up to fourteen feet wide in places, their fragments cemented with white, milky quartz. Besides the rusty iron reds, oxides of copper, malachite and azurite, spotted the outcrop in brilliant greens and blues. After exposing seams of galena and sporadic chalcopyrite with their hammers, the Niegolds quickly laid out their claims and took samples in to be assayed. Some of the samples must have assayed favorably in gold and silver since the Niegolds began to dig trenches, shafts and eventually tunnels to expose the ore bodies. Most newspaper reports mention that the Niegolds were obtaining predominantly galena and silver from their claims but assay maps also show that there were some good values in gold. Other newspaper reports mention that the Niegolds were obtaining their ore from pockets on the vein, which implies that not all of the vein was mineralized. The ore must not have lasted long because by 1884, the brother's Midland Mining Company was no longer shipping ore from Galena Mountain though

Precarious Perch

Judge N.C. Maxwell of San Juan County was the mine manager of the Old Hundred in the years 1904 to 1908. An interesting article by Jack Foster appeared in the Rocky Mountain News in 1953 telling of life at the mine in the early 1900's.

"The judge remembered the bitter weather and the blizzards that plagued the mine. There was really no place for the miners to walk. They had to jump like mountain goats from place to place. They could walk into the mine shaft (now sealed with snow and ice this year—1953) or into the tram house or to the johnny on the ledge outside. But that was all.

"He recalled that at one time there was a cook, a baker and two waiters caring for the needs of the crew and they could get food and water only from the valley below by ore buckets.

"Apparently in 1939, there was an effort made to do an exploration and sampling program. On July 14, a crew arrived, Jim Murphy was cook, but on August 17 they left leaving the note: 'This is hell. We started to heaven and landed square in hell.' And this was only summertime."

Real men worked at the Old Hundred and this was the crew in the fall of 1935. And their names are fading from the memories of the living… They worked the #2 level and either lived in town or in buildings by the mill. 1. Mill Foreman Hood 2. John Anzek 3. Amos Jaramillo 4. Slim Ruether 5. Tony Perez. 7. or is this Slim Ruether? 12. Dan Cummins 16. Louis Dalla 17. Jim Bell 18. Mr. Fedrizzi 21. Joe Skyzner 22. Steve Galinski 13. Lonnie Masden 24. Joe Minetti 25. Art Corazzo 26. Harry Turner 27. Jim Cole 28. Pete Dresback 31. Beans Lewis. *Mike Andreatta Collection*

they continued to prospect there for another fifteen years. Several geologic factors had soon come into light that influenced not only the Niegolds' mining methods but came to plague future mining by larger firms.

Probably one of the greatest factors in ore deposition in the San Juans, and especially influencing the #5 and #7 veins was that gold and silver often tend to be deposited higher in the veins while base metals such as lead and zinc are found in the lower parts of the vein. Good gold and silver values were found at high altitudes on the #7 vein above the boarding house but with depth the values were found to be negligible. One of the main reasons that the Old Hundred Gold Mining Company failed to find gold or silver on the #5 vein in their 1935–1936 operations was that the vein was eroded several hundred feet lower than the #7 vein and all of the high values were gone. Other operations on both veins, such as Dixilyn Mining Company's from 1967 through 1972, also failed to find any appreciable precious metal values at depth. Many mining operations have begun at lower elevations in the San Juans based on favorable assay data and old workings at high altitude, only to find that there are nominal values at depth and that the "old timers" had already mined out the good ore.

Another large influence on the formation of ore deposits is the availability of open space in a fracture for the migration upwards of hydrothermal waters or having open "plumbing." Hydrothermal waters, when encountering a barrier such as a narrow, choked fracture or clay seam, will migrate elsewhere. Earlier depositions of worthless quartz may have also formed barriers to later, metal-rich waters. This apparently happened on many several hundred-foot long sections of the #7 vein where the fractures are filled only with barren quartz. The sections that are mineralized are called ore shoots, and though tending to run vertically can also run diagonally through a vein. Miners on the #7 vein encountered this type of mineralization where better

Up at the Old Hundred Mine the boarding house door stands open around 1936. There is a deck all around this building and a connecting walkway to the tram house. Power lines are visible. In the tram house a window is open, and one can see an anvil, a saw horse and drill steel, and an ore cart. This is right at the mouth of the mine. The boarding house, at least, may have been cabled to the rock wall behind it at one time.
Paul Beaber Collection

values were confined to ore shoots and were found at high altitudes. During the later stages of gold deposition, much of the plumbing wasn't open or was already clogged with other minerals so gold mineralization was extremely sporadic and spotty, even within the gold ore shoots.

The Hawkeye tunnel was driven in the late 1880's on the northern extension of the #7 vein through several hundred feet of barren quartz. It was not only driven in a barren part of the vein, but also several hundred feet below where the highest precious metal values were found on the vein above. Very little stoping was done on the Hawkeye level, but directly above the end of this level, and several hundred feet higher, was an ore shoot that was mined along a length of 400 feet and up to 150 feet high. Some good values were found in this upper stope including a few samples that ran up to fifty ounces of gold to the ton but assays like these were few and far between, with a majority of samples yielding less than desirable results.

A majority of the good ores on the #7 vein were mined out by the Niegolds from the old #7 level (three hundred feet above the boarding house) to the surface. Huge gaping holes at the surface, where stopes were blasted completely through testify to the rich nature of the ores. Though a vein may be ten feet wide and completely filled with rock fragments and quartz, the metals may be confined to certain parts of the vein, such as the hangingwall or footwalls. The Niegolds learned this early on and may have confined their mining to narrower widths of two or three feet on the richest streaks within the vein. The Niegolds found a different kind of gold in the early 1900's with the promotion and sale of their Midland Mining Company's Old Hundred properties for four hundred thousand dollars in 1904.

"Reinhard, who, with his three brothers, had founded Niegoldstown beside Cunningham Creek at the base of Stony Pass back in 1876, was realizing his last fortune from mining. Niegoldstown had gotten a post office in 1878 and from mining claims on Green Mountain just past the Old Hundred, a wealth of silver had flowed to the brothers. The little town found itself with a piano and the four men from Germany would entertain their guest with operas done in

Party Favors

Miss Virginia Graham, 'Virgie' as she was known to her friends, once told me about going up to the Old Hundred boarding house on the tram with several other Silverton girls to attend a dancing party. This would have been around the turn of the century. A number of college boys were working at the mine, and they had planned quite an elaborate affair. She remembered that each young lady was given a pretty china cup and saucer as a party favor. It seems that the mine boarding house was using a particular brand of coffee, each package of which contained the china as prize. The resourceful young men had saved enough for each girl to have one.

—*From Ruth Rathmell,* Silverton Memories

Top floor, the shifter's room, 1994. The boarding house was last used full time in 1908. There were roughly 25 bunks upstairs. David Emory Photo

The Old Hundred kitchen, 1994. A large chopping block is on the right. The bench probably was originally in the dining room. David Emory Photo

costumes of knickerbocker pants, powdered wigs and buckled shoes. Turkish tobacco at five dollars a pound and fine wines were consumed by the Niegoldstown founders. But by 1884, avalanches had destroyed the last of the settlement and the fortunes were gone. And though Reinhard had stayed on in Cunningham until the sale of his Old Hundred properties twenty years later, he had little time to enjoy his last fortune realized from this sale, dying a short two years later. Today, there is no trace of Niegoldstown and only the Old Hundred lives on as a mining tour."*

The Old Hundred Mining Company was formed and did most of the extensive development found on the two big veins from 1904 to 1908. They were responsible for building the massive boarding and tram houses far above Cunningham Gulch, along with three tramways to connect the various levels of the mine. They drove several thousand feet of tunnels and did a majority of stoping on two ore shoots above the new #7 level, one to the south of the seven level crosscut and one to the north.

The Old Hundred Mining Company had driven the new #7 level three hundred feet below the Niegolds' old level in the hopes of encountering the same rich ores, but found that the values had fallen dramatically between the two levels. The fall in values that the company encountered may not only have been due to zoning but also to dilution caused by wider mining widths or taking the vein in its entirety. It was often a common mistake for mine superintendents, engineers or geologists to take samples only on the richest parts of the veins where there was obvious mineralization, instead of the entire vein, and then be surprised when the actual mining widths (usually wider than the width of the samples cut on the vein) produced much poorer values.

Often there are parallel fractures to the main hangingwall of the vein, and as soon as the vein is mined, these fractures break and add waste rock to the stope materials. This was probably a factor on the #7 vein and especially on the #5 vein where some stopes near the surface were taken to the full width of the vein. Erosion was also a factor in mining on northern parts of the #7 vein, as most of the good gold values evident at higher elevations to the south were eroded off to the north.

* *From Allen Nossaman,* Many More Mountains, Volume II)

The company also ran into other troubles on #7 level as they drove the tunnel to the southeast. Several east-west trending faults were encountered that offset the big vein to the west, leaving miners and engineers alike in a quandary as to which direction to follow. The vein appears to have been found after each offset, but there remains a high possibility that a parallel, mineralized fracture may have been followed instead, and the main vein still remains somewhere nearby.

The Old Hundred Mining Company failed after only five years of mining activity, spending over a million dollars with returns of less than a half million. Not only had the geology been a factor but there were also human factors as well. Mishler wrote in 1931 that of the $500,000 dollars spent on actual mining operations, $275,000 dollars went to promotions and deep pockets.

The mine sat dormant until bought by Carl Dresser in 1927, who died shortly after. Mrs. Dresser sold the property to a group of Pennsylvania investors in 1934 who then began mining operations on the #5 vein in November of 1935. The #5 vein was chosen because it was more accessible than the #7 vein. It was hoped that the proceeds from mining on the #5 vein would be used to drive a tunnel from #2 level back to the #7 vein but driving long tunnels takes money, which they didn't have. The operation closed in March of 1936, having made five thousand dollars on ore milled after expenditures of $122,000 dollars.

The Old Hundred tram house, 1994.

David Emory Photo

The operation was hampered by severe weather, ice in the old workings, and unexpected costs in revamping a mine that had sat dormant for almost thirty years. They had also made the mistake of spending a lot of money fixing up the old mill and tramway, adding equipment, and building structures before enough ore was demonstrated to exist to warrant the costs. This seems to have been the most common mistake to occur in the San Juans, not only ninety years ago, but even in the 1980's. Seems like folks always forget to prove that there's any ore!

The geology of Galena Mountain soon began to have its effect upon the company. They hadn't realized that the gold and silver deposited high in the vein were long gone due to erosion and the only grades of ore left at lower elevations were too low to mine. One other theory is that the main episodes of quartz mineralization blocked off the vein so completely that later gold-laden waters could not migrate up the #5 vein at all. Apparently, the veins were not open or were not in the right vicinity for heavier lead/zinc mineralization such as that found further south in Cunningham Gulch at the Osceola, Green Mountain and Gary Owen mines.

Whatever the reasons, the Old Hundred Gold Mining Company's activities ground to a halt, and the mine essentially sat dormant for another thirty years.

Geology again came to be a large factor in the failure of the next company to try operations at the Old Hundred mine. Dixilyn Mining Company came into the San Juans in the late sixties with lots of money and left without it in 1972. Dixilyn began an extensive diamond drilling program to find ore potential on the two big veins. After two big gold intersections near the #7 level, the company began driving tunnels on one and two levels towards the #7 vein, while driving levels and raises on the #5 vein. An Alimak raise was driven upwards to one of the gold intersections but little gold was ever found, even after the original exploratory drill hole was actually found. Two ideas exist as to this discrepancy in gold values. Some think that the drill core may have been salted, while others believe that the Old Hundred or #7 vein had very spotty gold, and high gold intersections by drilling were definitely possible. Even if there had been good gold discovered in the drillholes, there was too much low grade ore in between. Again, zoning became the predominant factor, as it was later discovered that gold values at lower depths were negligible at the intersection of one and two levels with the #7 vein, hundreds of feet below the #7 level.

The low assays on the vein at #1 level were blamed on a poorly mineralized shear zone within the vein. Realistically, the gold and silver were probably not there, neither on the #7 vein or the #5 vein. Millions of dollars were spent on exploration and development by Dixilyn Mining Company but little ore of value was ever taken out of the lowest workings of the Old Hundred Mine.

Another large mining company came into Cunningham Gulch in the early 1980's to explore, sample and drill the Old Hundred property in the middle of a San Juan winter (most companies do exploration in the summer). One story probably sums up the results.

The company hired a geologist who had never logged drill core or worked underground before in his life, to oversee their drilling program. Normally, a geologist will try to intercept a vein with a drill hole by drilling perpendicular to the vein to obtain the optimum intersection. After looking at Dixilyn's hot gold intercepts near the #7 level (the fateful X in the mountain), the geologist set up his drills to get as close to this location as possible and save the costs of lengthier drilling footage. Unfortunately, the drill holes were positioned parallel to the vein instead of perpendicular to it with the possibility that a hundred feet of vein width would appear to have been drilled, even though the vein was probably only ten feet wide.

Timing was bad. The mistake was realized after excited phone calls to the main office were made about a big quartz intercept. Wrong! It was discovered by the core loggers that the drill had been reaming a one inch wide quartz stringer along the fault line for forty or fifty feet. The company and their geologist left with the spring thaw.

And so it goes—the Old Hundred Mine is now the Old Hundred Mine Tour as tourists look in awe at the old workings and dream of the riches untouched by the old timers. Dreams are only dreams though and not the reality of geology. I've got two shops in town now. I'll be busy mining another type of gold come summertime. These veins haven't played out yet. ❋

—Contributed by Scott Fetchenhier

Scott Fetchenhier, a miner at heart, contributed this article. Scott, a geology graduate from Fort Lewis College in Durango, has lectured on rocks to the children in the area and has taken them on field trips. Although two shops in the town keep him busy, he still finds plenty of time for his other loves– kids and diving into holes, hoping to find old workings to explore. John Marshall Photo

The Old Hundred boarding house. Silverton can be seen in the distance. **David Emory Photo**

BUNKS

KITCHEN

BUNKS

DINING ROOM

TOP FLOOR

MAIN LEVEL

David Emory Photos

The Handlebar Restaurant Collection

Have a quick look inside the tram house to make sure everything is in order. Too bad there's not an empty bucket headed down. The trammer could grip the bucket out and we'd be off, over that first tower out, back to Cunningham Gulch far below. But first, we'll do some more exploring up toward the Highland Mary. See the road through the glassless window, the one that switches back? Stony Pass. The straight one goes right up to the Highland Mary. Now, some agile footwork, a jeep, a little luck, then maybe we'll be down in the valley. Since it's 1994, I guess we'll have to skip the tram. It must have been one incredible ride.

Warren Prosser Photo, Bud Frizzell Collection through Zeke Zanoni

The rails end at the Green Mountain Mill but the mines and mills continued on up the valley. The next structure up Cunningham Gulch was the Vertex Mill then the Buffalo Boy tramhouse. From there, we continue up to the Highland Mary Mill.

Andy Hanahan Collection

John Marshall Photo

(above) The Silverton town fire truck, a 1929 LaFrance, was sent to the Vertex fire of 1935 to no avail (left). Some sixty years later, November, 1992, the little fire truck was itself rescued from a fire that nearly destroyed it and the town hall.

The upper Buffalo Boy tram house and interior workings still stand in a basin at the side of Galena Mountain near 13,000 feet while far below in Cunningham Gulch the lower tramhouse also sits idle, geraniums in the windows, 1994.

John Marshall Photos

The Highland Mary Mill was located at the end of Cunningham Gulch. We have left the train and followed the road seen so clearly from the Old Hundred tramhouse on page 90. We travelled along the side of Green Mountain, past the Osceola Mine and the Pride of the West Mine, under the tram cables of the Little Fannie Mine, past the cribbing and that old rusted truck body by the Green Mountain Mine right on to the Highland Mary.
Mike Andreatta Collection, 1950

THE HIGHLAND MARY

Today you can find the stone foundations and some of the timbers but most is gone. The Highland Mary Mill had stood for over seventy-five years until 1952 when an avalanche struck, partially destroyed it, and started a fire. The watchman was killed and these buildings were destroyed. The combined disaster effectively ended operations, 1950 photo. *Mike Andreatta Collection*

Mining on Orders from a Spirit!

This mine not only had an early start but a quite unusual one. The Highland Mary was developed by Edward Innis, a New York capitalist described by a newspaper of the period as "so eccentric that San Juan people regarded him as a crazy man." Innis left New York in the spring of 1875, claiming that the "spirit" had directed him to go over the mountains and plains from New York until he would find a point from which he could drive a tunnel into a lake of silver.

He traveled to Pueblo, as far as he could go by train, then outfitted a tremendously expensive pack train and headed toward Silverton.

He had with him a chart made out by the seer, showing the route to be followed by the pack train. Near the head of Cunningham Gulch, he told his men: 'Our journey is ended. This is the place I have been hunting.' That was in July, 1875.

The place he chose was marked by the remains of two shacks left there two years before by some prospectors, J.C. Dunn and William Quinn. Innis located Dunn and Quinn and bought their twelve claims for $30,000 dollars. He then located several more adjoining claims, gathering a total of seventeen.

Assays on each of the claims ran high. Innis was pleased, but said these were only incidental to the "lake of silver" he would find inside the mountain.

Innis was financing his project on the interest from $1 million dollars he had invested in the East. The following spring he built engine houses, a boarding house, a $10,000 dollar mansion for himself and brought in machinery which cost thousands in transportation alone.

At the end of the season, he loaded one large shipment of ore that averaged $7,500 dollars a ton. Most of the valuable ore taken out was thrown on the dump.

Innis returned East and did not come back until August 1877, when he began work on the Highland Mary's tunnel called 'the most remarkable in all the annals of mining in the world.'

Every inch driven was done on explicit orders from Innis, and he maintained that these orders came from the 'spirit.' According to reports, every time the course was changed on spirit orders, they struck a rich lead, but the erratic policy gave the big bore an irregular course unduplicated anywhere.

Because established rules of mining were violated following the 'spirit,' a number of men were killed as the work progressed. The men, however, began to believe in the 'spirit' and refused to enter the mine alone.

Once Innis telegraphed from New York: 'Turn to the right. Follow that crevice.' The mystery was how Innis knew there was a crevice when he had been thousands of miles away from the work for weeks.

An old miner also recalled, 'In the summer of 1879 we got a telegram from Innis telling us to look out for water. We took it lightly, but in a few days, our shots broke into a cave of water. We jumped in a car and rode to safety, but it took four days and nights for that subterranean lake to empty itself.'

On spirit orders, Innis continued to 'drive ahead,' shipping none of the rich ores located in the work. By 1885, he had put $750,000 dollars into the mine. The tunnel was a mile long, with leads going to 2,700 feet below the surface.

Innis' money was invested in the New York banking house operated by former President U.S. Grant and Ward. The bank failed in 1885 and Innis lost his fortune.

He owed $12,500 dollars on mining expenses and asked his creditors to accept his notes. But the creditors wanted security on the mine itself, which Innis' 'spirit' ordered him not to give. Innis returned East and died insane.

The Highland Mary property was entangled among a number of Innis' heirs, and these refused to cooperate with other promoters who wanted to take the mine. As a consequence, the mine lay idle for fifteen years.

Then Mrs. Mary B. Murrell, widow of a St. Louis physician, heard of it. Mrs. Murrell finally got rights to the mine from all the Innis heirs, and paid the fifteen years in back taxes on it. She interested a number of wealthy Easterners in the mine and formed the Gold Tunnel & Railway Company in New York.

The Highland Mary's operation was reinstituted in February 1901 under Mrs. Murrell's personal supervision. Almost immediately profitable shipping was begun. By 1907 the Highland Mary was the second largest producer in the Silverton area. There is no record that the 'lake of silver' ever was located. ❋

—From the Rocky Mountain News, 1953.

The White House at the Highland Mary was built in 1875 and burned to the ground in 1942. The first wedding in the new San Juan County, previously known as La Plata County, took place here in August, 1876. Edward Innis, who created the original mine just below here, used the building as his residence and office. In 1878 the edifice became a post office and Innis became a postmaster. The #7 level was above here, the #8 level lay below where foundations can be seen today. The White House stood close to the junction of where a road comes up the hill straight from the mill. Today this road is blocked from joining the jeep road one drives on to get to the Highland Mary Lakes trailhead. There were basically three phases to the development of the mine; the Innis period, from 1875 until he returned east and went insane around 1885; the Gold Tunnel and Railway Company period which operated from 1900 to around 1925; and the Joe Bradley period from 1940 to 1950. Mike Andreatta Collection

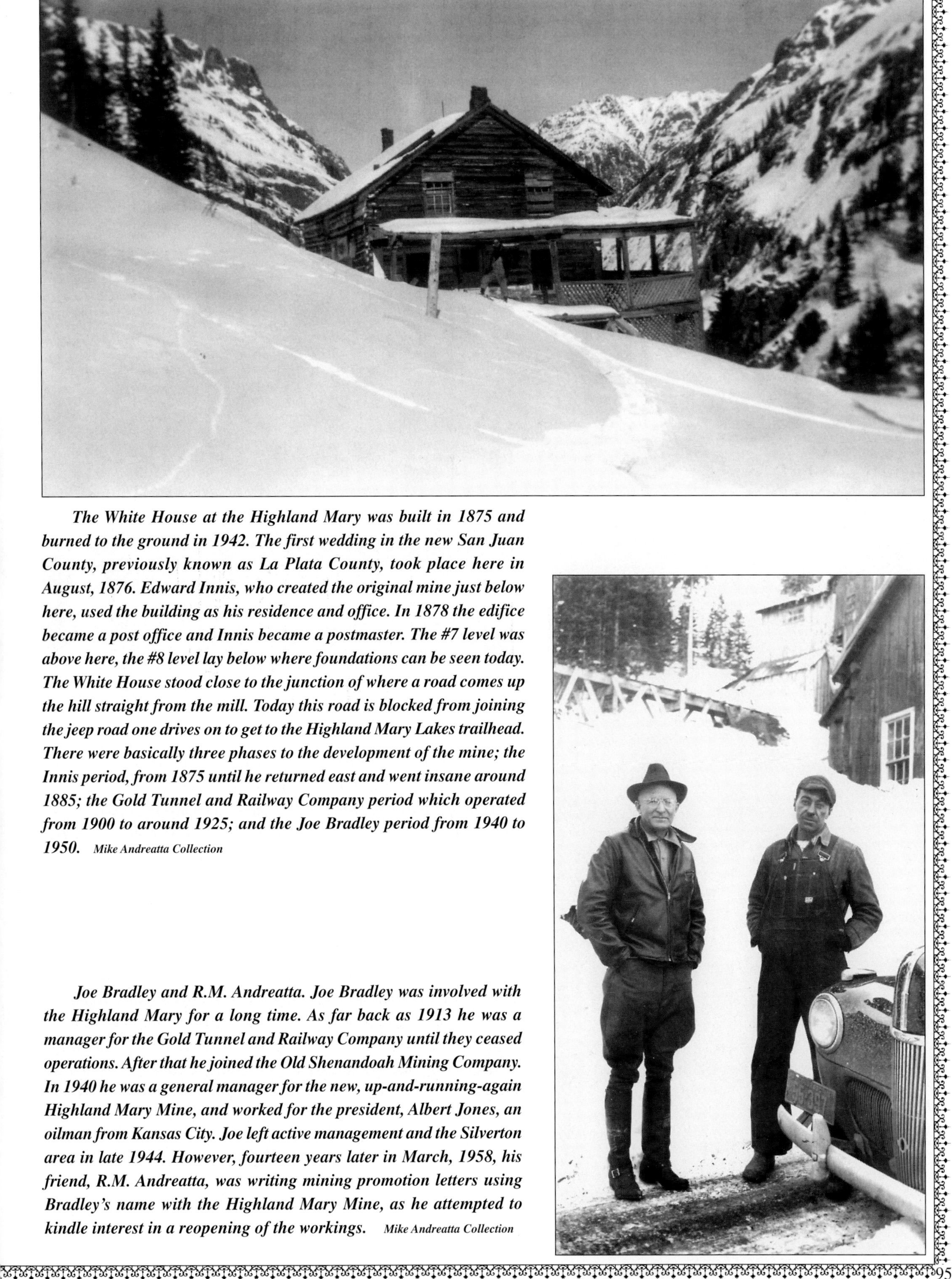

Joe Bradley and R.M. Andreatta. Joe Bradley was involved with the Highland Mary for a long time. As far back as 1913 he was a manager for the Gold Tunnel and Railway Company until they ceased operations. After that he joined the Old Shenandoah Mining Company. In 1940 he was a general manager for the new, up-and-running-again Highland Mary Mine, and worked for the president, Albert Jones, an oilman from Kansas City. Joe left active management and the Silverton area in late 1944. However, fourteen years later in March, 1958, his friend, R.M. Andreatta, was writing mining promotion letters using Bradley's name with the Highland Mary Mine, as he attempted to kindle interest in a reopening of the workings. Mike Andreatta Collection

(above) *The Highland Mary Mine, #7 level, about 1905. Today the Highland Mary trailhead starts just below here and to the left. Spencer basin is up behind the buildings. This site was abandoned in the twenties. The large building on the left was the stables and blacksmith shop. Just to the right of the blacksmith shop stood the portal. In the foreground stands the ore house and the big building is the boarding house.*

What's in a picture? Usually more than meets the eye. Warren Prosser, the photographer, was born in Silverton in 1875 and became a consulting mining engineer—and apparently also a photographer. He owned and operated mines in the area as well. In the twenties he was a representative for AS&R, in charge of leasing the Silver Lake Group. Zeke Zanoni's grandfather, Ernest, leased some claims from Prosser.

Then in the seventies, Bud Frizzell showed up in Silverton. (Prosser died in 1975 in Denver.) Bud had acquired several glass plate negatives, mining information, and more from Prosser in Denver. Bud was a mine promoter and bought the old Root and Norton assay office. Meanwhile, Zeke and his brother Tom were running the San Juan Museum on Mining History located diagonally across the street from the Grand Imperial Hotel (where Danny Tuma has his shop now). Bud gave prints to the museum. When Bud died in the eighties, his daughter gave the plates to the San Juan County Historical Society. The prints here came from Zeke Zanoni. Silver threads that wind in and out…

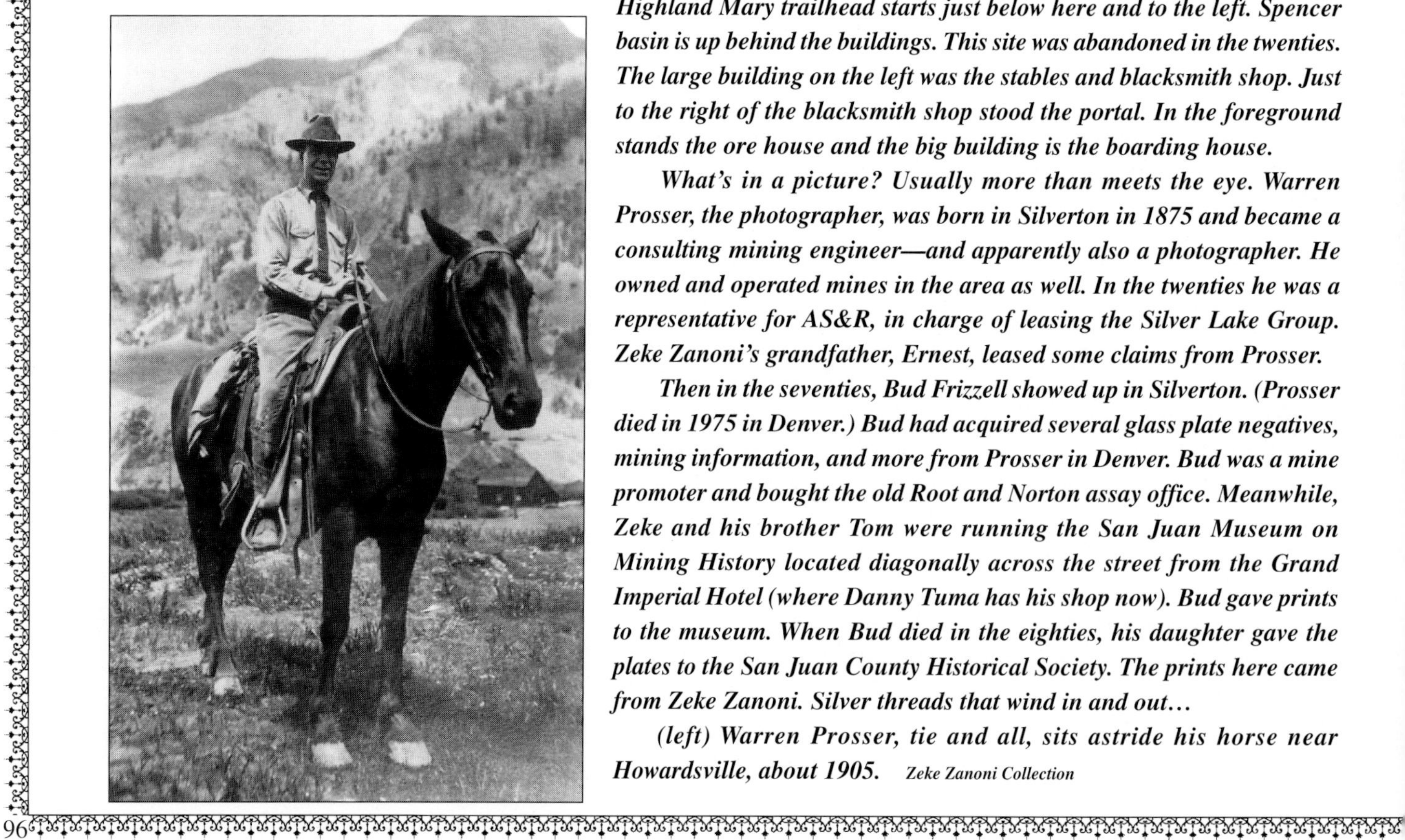

(left) Warren Prosser, tie and all, sits astride his horse near Howardsville, about 1905. *Zeke Zanoni Collection*

This is the #1 level (the top level in this case) of the Highland Mary Mine, by the present day Spencer Basin trailhead. Bob Sturdevant, Claude Robinson, Dave McClure, Jr., Charlie Hawkins, Frank Hitti, and Mike Musich pause in the daylight, 1945. *Mike Andreatta Collection*

Operations at the Highland Mary

Most of the highly technical material that follows about the Highland Mary operations has been taken from articles appearing in several mining magazines between 1946 and 1950. Of interest are the sophisticated engineering methods employed to provide power (some installations were in place by

Installation of the water line from the Highland Mary Lakes to the mill is put on hold so a picture can be taken. That's a welder's oxygen tank on the ground. *Mike Andreatta Collection*

R.M. Andreatta, Babe Sturdevant and Bob Sturdevant. Babe and Bob bought the pneumatic hammer, a three foot shell drill because they thought they could go through the ground a lot faster. It was air driven with a water head. The mine saw the light and slowly paid off the cost of the drill back to the miners. After all, steam drilling had first come to the San Juans at the Highland Mary in 1876, starting the end of double-jacking. Note the box of powder on the ground, 1946. In 1946 some of the guys made $4 dollars a day. The elite, the miners themselves, made $6 dollars an hour. *Mike Andreatta Collection*

1906) operating right next to raw horse—or should we say mule—power.

"The Highland Mary is located at the head of Cunningham Gulch some 8° miles from the town of Silverton. The mill is on the floor of the scenic valley and the mine portals are on the rocky hill above. A 1,200 ft. aerial tramway provides transportation between them. Although at some 10,000 feet of elevation, the location is a good one. There is plenty of water and water power.

WATER FOR ELECTRIC POWER is taken from the Highland Mary Lake at the head of Cunningham Creek and piped some 7,000 feet to the mill (see photo page 97). Collected in a fourteen inch pipe which reduces to a twelve inch and again, just above the Pelton wheel, to five inches, the water is delivered through a 1$^1/_8$ inch to a 1$^3/_8$ inch nozzle. With an 832 ft. head, this provides about 360 psi. at the Pelton (water) wheel. The installation was originally made in 1906. The thirty-six inch single nozzle Pelton wheel, equipped with a Woodward governor, drives a 250 h.p., three-phase, sixty cycle synchronous motor, and delivers power at 440 volts. Another thirty-six inch Pelton wheel furnishes power for the operation of an Ingersoll Rand Imperial Type 10,450 cu. ft. compressor, backed up by a seventy-five horsepower motor, that supplies the mine with compressed air. Also a Chicago Pneumatic 320 cu. ft. air cooled compressor is available for additional air at the mill or mine. Utility power makes up any difference in load and also furnishes control.

ELECTRIC POWER is furnished by the Western Colorado Power Company. Due to slides, mine operation is April – December. Higher power costs result due to this not being a year-round operation. Also, slides destroyed the power line each year. It was impossible to keep a pole line in the gulch on account of snow slides so Hammond Matthews designed a suspended line that has been free of trouble from any source. Three cables were strung across Cunningham Gulch from points above the slide-forming areas. The 3-wire transmission line is suspended high above the valley floor by these cables, thus eliminating power difficulties due to slides. It has four spans from 1,405 to 1,760 feet long of No.4 stranded steel wire, 5,600 lbs. MBS, these spans being suspended from three cross cables anchored in the cliffs on each side of the canyon. The transmission wires are fifteen feet apart and are from 30 to 350 feet above the ground. *(These lines were removed forever from the valley in the summer of 1995.)*

In 1949 the main working level is entered by way of the Bradley Crosscut at an elevation of 11,350 feet.

TRAMMING OF THE ORE TO THE PORTAL is about 3,000 feet and by mule *(in 1946!)*. Card Timken Bearing 20 cu. ft. ore cars are used. Seventy to eighty-five cars per shift are trammed. A mule could pull a five-car string of cars over the 18" gauge track of 30 lb. rail. However, a General Electric two-ton locomotive with Edison storage batteries was on hand and ready to go into service as soon as the erection of a battery charging station could be completed.

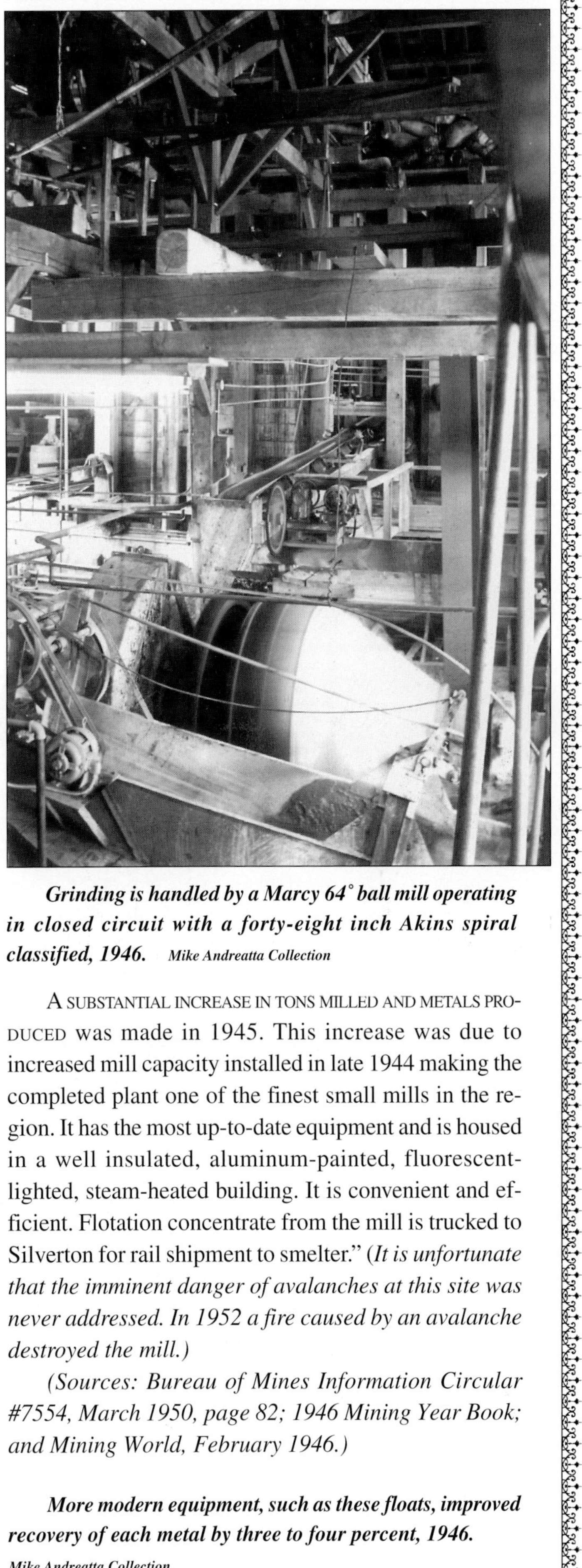

Grinding is handled by a Marcy 64° ball mill operating in closed circuit with a forty-eight inch Akins spiral classified, 1946. *Mike Andreatta Collection*

A SUBSTANTIAL INCREASE IN TONS MILLED AND METALS PRODUCED was made in 1945. This increase was due to increased mill capacity installed in late 1944 making the completed plant one of the finest small mills in the region. It has the most up-to-date equipment and is housed in a well insulated, aluminum-painted, fluorescent-lighted, steam-heated building. It is convenient and efficient. Flotation concentrate from the mill is trucked to Silverton for rail shipment to smelter." *(It is unfortunate that the imminent danger of avalanches at this site was never addressed. In 1952 a fire caused by an avalanche destroyed the mill.)*

(Sources: Bureau of Mines Information Circular #7554, March 1950, page 82; 1946 Mining Year Book; and Mining World, February 1946.)

More modern equipment, such as these floats, improved recovery of each metal by three to four percent, 1946.
Mike Andreatta Collection

A timber car on the jig-back tram: 1,200 feet long, two minutes time, 650 feet in elevation change. Ride at your own risk. Mike Andreatta Collection

The tram, a 'jig-back' type, was used for transportation of all personnel and supplies to the mine. A jig-back is a two-bucket tram system usually used for covering short distances and relies on the weight and motion of one bucket coming down to pull the other bucket up. In a sophisticated system such as the Highland Mary, a motor at the mine level applied power to the bull wheel to pull a load of timber and men up the tram without relying on the counterweight. "Bucket capacity is 1,300 pounds of ore and tramway is operated by fifteen horsepower motor with V-belt drive and bevel gears to grip sheave. Motor has magnetic brake for braking when power is off; when motor is operating loaded bucket is braked by regenerative motor braking. Buckets are dumped by automatic dumping device into hopper and then by sixteen inch belt to 250 ton crushed ore bin. Telephone communication between loading and dumping stations is used for clearing." **—Mining World, February, 1946.**

The Tram Never Slowed

On the afternoon of November 13th, five days short of his twenty-fourth birthday, Douglas Caine had gone down to the mill for parts to fix an airhose. While there, the big young man stopped to talk with Joe Todeschi. Joe gave him some bits for the drills up above. "But there's no need to go back up now," says Joe. "It's about end of shift."

"No," the kid replies, "the miners need them."

Now, Henry Grey was running the tram at the top. No one was working it down below. Doug called up on the phone and told Henry he was getting on and to bring him up. Doug had worked at the Highland Mary before. He'd taken a short vacation to visit some other western states. He'd only been back to work one day. When he climbed into the self-dumping buckets, he neglected to

Tramway cables stretch to the Highland Mary Mill, 1950. This spot is very close to where Doug Caine fell to his death.

Mike Andreatta Collection

slip the locking ring down over the arm that would keep the bucket from dumping him. The tram took off. The bucket dumped. Doug made a quick decision. The wrong one. And it cost him his life. Instead of letting go and falling a few feet, he wedged his body in, bracing himself with his feet. The tram never slowed. Harry Hoffman was working on the tailings pile. He heard a man yelling. Just about that time the bucket banged against the splice on the stationary cable. Doug hadn't counted on that. Harry saw him fall to the rocks and snow far below. Harry ran to get Joe in the blacksmith shop at the mill. When Joe figured out what Harry was saying, he went and got John Troglia, the mill operator. They climbed up to Doug. At this point, no one could have helped the young man, but they tried. They got the stretcher of two by fours and burlap sacks and slid and carried him to the bed of a pickup truck parked by the mill. John drove and Joe rode in back with a bundle of empty hopes. They sped into town and Rosy Giacomelli met them at the hospital door. Bill Maguire, Sr., the undertaker, heard the noise and came to take a look. "Boys, this man is dead. There's no need to bring him inside." Bob Caine, who worked the shift opposite his brother, was also at the hospital. "Oh my God! Would you boys tell our brother Jack? He's out working the county road with the D-7." Joe and John slowly went back to the Highland Mary, silent. They found Doug Caine's hat on the ground. Just outside the tram house. ❋

A pine marten hangs out in some mine cribbing, 1994. Jess Carey kept its ancestors happy with good food. In return they kept him company at his lonely outpost. *John Marshall Photo*

The Highland Mary reopened in 1940 for the war effort and remained open until 1950. For some of those years Jess was the caretaker. In those years the mine only worked from April or May until November or December, depending on the severity of the winter. Then conditions were deemed too dangerous and the mine would close, leaving Jess alone at the end of the valley for the harshest months of winter. Not completely alone.

WINTER SOLITUDE

Amazingly enough, Herman Dalla, working for the county, would periodically show up in a plow or cat to maintain the road. Jess could hear his engine echo up the valley in plenty of time to bake an enticing snack to lure Herman in for some conversation. Jess had cooked at a number of mines around the area. Herman made welcome company. Human visitors seldom came that far. Between visits Jess would have to settle for one way chats with the resident pine marten, who in turn were only too happy to put up with any gossip in trade for some of those fine kitchen fixings. ❋

Dinner is held on the second floor of the Masonic Hall on 13th and Reese Street. The structure was built in the 1880's and stands intact today. Here, dinner is being served in 1947. In some way every one of these people's lives were connected with the mines in the hills surrounding Silverton. The wives were well aware of the dangers many of their husbands and sons faced every day in the countryside around this mountain town.

Jess Carey, the first man on the left in the picture, was one who had worn many different hats in a varied career. As he grew older his jobs became quieter in nature.

1. Effie Andreatta 2. Frances Beaber 3. Jess Carey 4. Bill Steadman and Mrs. Steadman 5. Gus Olson and Mrs. Olson 6. Oly and Mary Rhoades 7. Hank Erlinger, mine promoter 8. Ed Koltz 9. John Arietta *Mike Andreatta Collection*

After a few winters at the Highland Mary, even Jess Carey deemed it too rough and too lonely a place to winter. For a while John Tomseck came to take over the watchman's duties. John had been in that line of work at the Sunnyside but even with that background John didn't stay long. At the onset of winter, 1951, John heard a voice telling him to get out: 'Leave John. Don't stay here.' John listened and left.

Danger was always a companion

The big winter storms of '51 and '52 came in strong. Fire insurance dictated that mines should have a caretaker. Somehow they found Harrison Stoker McCormick in the bar at the Grand Imperial. (Dead drunk? The records don't say, but it would have been better.) Facts say he never should have left town. Yet right around Christmas he reached the Highland Mary just in front of a real San Juaner. During the fierce storm around January first a huge slide roared through the trees down onto the mill and Stoker McCormick. The slide had never run before, and later, rescuers would claim the snow to be two hundred feet deep in places. One of the oldest mills left in the country was wiped out. Damage was $250,000 dollars. Stoker's quarters were below the mill and he was in bed, a crude affair of two by twelve mine lagging, when the slide hit. The six by six ceiling timbers above him—the mill's floor—snapped and broke loose pinning the poor man to his bunk. The stove was knocked over and the building began to burn; what wasn't buried by snow was consumed by fire. Rescuers couldn't tell for sure which was the killer, but the watchman's life was surely gone. The mill was never rebuilt and the mine never ran again.

Jerry Sandell was living with his father in Cunningham at that time (see their house on page 81). Jerry had been partly raised in Cunningham and had loved every minute of it. He'd fished the beaver ponds, the same as the ones there today. But to see first hand the havoc wrought by nature's fury was hard on the young boy. Men at the Pride of the West Mine could see smoke at the end of the valley. A rescue party was formed. Eleven men left including Bill Loftus, mine foreman, and others like Bud Crawford. Since Jerry's father, Glenn, was going, they let Jerry come along. They slowly made their way to the mill. The twisted, smoldering wreckage was so tossed about that getting to where Stoker's room had been was fairly easy. There the tortured body lay, pinned to his bunk, flashlight in hand. Bud Crawford helped stick the remains in a bag. (Bud had cleaned up beaches all along the Pacific during World War II.) A little lunch and they'd all be headed back. Bud made himself comfortable on top of that body, neatly wrapped in a bag and the only dry place to sit, and proceeded to enjoy his lunch. Jerry looked at the sled, the body, then Bud. He couldn't even think of eating. In due time, Harrison Stoker McCormick had his first and only ride down Cunningham. Last stop, the Hillside Cemetery. And John Tomseck, the caretaker he'd replaced? He joined Harrison at the cemetery after kidney complications. One died the first day of that year, 1952, the other on the last. ❋

Jerry Sandell heading up to Kite Lake in 1990.

Steve Smith Photo

John McNamara Collection

There was a web of life strung between Silverton and the surrounding towns and mines and strands through time as well. Jess Carey, was born in 1888 and spent a good deal of his life around Silverton. At one stretch in his later life he ran a museum on Blair Street where Zhivago's stands today. John Troglia, who, in a couple of stories back, had tried to rescue Doug Caine, owned the building. On display in the museum was the bar pictured below. It had come from the Zanoni-Padroni saloon a few blocks down the street. That saloon would eventually become the Bent Elbow restaurant. The Bent Elbow was run by R.M. Andreatta, who previously ran the Highland Mary Mine. When the museum closed the bar was moved to the lobby of the Grand Imperial Hotel. And you can go there today, rest your elbows and swap lies with anyone nearby, like many before you.

Andy Hanahan Collection

IT WAS THE BRIGHT LIGHTS OF TOWN THAT WOULD DRAW THE MEN FROM THE MINES. Town was where stories were told and history wove a web. People would spend weeks even months working in the mines in the hills around town. Then the magnet of warmth, light, and the need for interaction with other people would draw them into Silverton. It could work the other way too. Town would get too small, too boring, and they would return to the mountains, the mines and their work. But people would know each other and be involved with each other. It was a web of life.

There were bars, bars with women, women to dance with and more. Bars with action, excitement, fights, open twenty-four hours a day—year round. The men used them as a release from the dangers of their job. The bosses were glad they were there for the workers.

There was a social structure in town and the women of Blair Street knew their place. But even the most "respectable" of citizens could forget where that line was and find themselves in the midst of the excitement of the nightlife of Silverton.

And as with any place that had hard and dangerous work around, the excitement of release was momentary and short lived. The lucky ones would settle down to a wife and a more normal job like restaurateur or merchant. Those who didn't often met sadness and misfortune. The ministers and priests had plenty to talk about. And the cemetery on the hillside above town holds almost three thousand graves. ❋

Her name was Nigger Lola. Her profession was the world's oldest and she was one of the last to ply her trade on the dirt street called Blair. *Andy Hanahan Collection*

The cards are dealt, the money's on the table and beer is offered, 1929. In the Laundry. Now why would a gambling establishment be called that? Jack Gilheany and Carl Longstrom's father Charlie owned the establishment located at one time in the basement by "Fetch's" on Greene Street. That's Charlie on the right with the flamboyant hat and cigar. *Edwin Larson Photo*

An illuminated raise in the Old Hundred, 1994. **John Marshall Photo**

The mines and the miners were a steady source of stories, sometimes songs, and even poetry. Some were of men who were amazingly lucky, others were of death—the unlucky ones. Yet in spite of the friction caused by close quarters and hard work, more often than not a strong camaraderie would develop between the men and later, the women, too.

Sullie's Pail

I've a thing or two to tell you that I think you ought to know
About that rusty bucket Sullie carries down below.
You are not the first one, stranger, to have laughed at Sullie's pail.
You're the only one that's laughin' now, the rest have heard this tale!

Well, now, when we were young and handsome had some ten years in the game,
Old Sul, he had a partner and Jim Reilly was his name.
They'd banged around together—Bingham Butte and Co'erdelaine
And brawled in every barroom from Ely to Fort McClean

How me and old Ted Johnson, sure you'll not remember him,
We were workin' in the raise, had a stope with Sul and Jim;
The four of us together we were workin' side by side
That's how come I chanced to be there on the night Jim Reilly died.

Well the blastin' had been easy, it was comin' out like sand
And we was muckin' out the ore—those days we mucked by hand.
We was nearly finished and I hadn't heard a sound
But something must have happened for Jim Reilly yelled bad ground.

When we headed for the timber Sul he must have took a spill
For when we looked back in there he was pinned beneath his drill
The ceilin' it was groanin' now all set to drop the lid
And Sullie pinned beneath his drill was sobbin' like a kid.

Now there's men who can watch their partners die and not throw their lives away,
But Reilly wasn't one of them, he wasn't built that way.
As soon as he saw what happened, "Hey hold on there Sul," he cried,
And before he had the words out he had thrown the drill aside.

Well they came around an ore car Reilly wearin' a big grin
Guess he never knew what happened when the hangin' wall caved in.
Sullie reached the timber and his face was white as chalk.
Reilly, ten yards back of him, had caught fifteen ton of rock.

Sullie's pail was buried; he ate from Reilly's pail in tears,
And he's carried that same bucket for more than twenty years.
So you can laugh at Sul because he's mean and drinks a lot.
But don't laugh at Sullie's bucket, it's the only friend he's got.

Author: Don Gibbons
Recorded: Tom Paxton- *Ain't That News,* Electra Records

The following story, written by Thomas J. Trumbull, first appeared in the Colorado State Historical Society magazine, Colorado Heritage, *in autumn of 1994. Before that printing*

Single-Jacking Fool for a Dollar a Day

the author unfortunately passed away and never saw the article. It is with the kind permission of his nephew, Tom Lindquist and his wife Blanche, trustees of Trumbull's estate, that this article can be reprinted. It so captured the atmosphere of what went on in the hills around Silverton over the years that this inclusion proved irresistible. So far the mines we have talked about and explored were the large mines—the company mines. For every one of those mines there were a dozen or more that existed like this one described in the following story.

In the depths of the 1930's Depression it got to where the offer of a job, any job, looked good, and one must keep that in mind when reading what I did in the autumn of '31.

I was offered a job as an apprentice miner, to work at a mine northeast of Silverton, Colorado. Wages a dollar a day, board and room paid. I was told that the work would be hand mining which means working without machine help of any kind, but nothing was said of the location of the mine with respect to the town of Silverton, and unfortunately it didn't occur to me that I should ask about that. Two others, students at the same mining school with me (the Colorado School of Mines), also temporarily out of school for lack of funds, accepted this offer and on November 16 we met at the appointed place in Denver, each with his bedroll and suitcase.

Thomas J. Trumbull

Our new boss met us with a shiny prosperous-looking stake-body truck. We loaded the truck with picks, shovels, roofing paper, dynamite, canned goods by the case, and such. We threw our gear on board, climbed on and headed for Silverton in the high country of southwest Colorado and in spite of an early start from Denver, it was late afternoon and snowing by the time we reached Montrose. All agreed it would be better to continue than to pay money for a room for the night. We were four; three riding in the truck cab, with the fourth on top of the load in the rear.

As the evening wore on, this outside spot got rougher and we could leave a man up there only a short time. It was midnight when we reached Ouray, which sits in a deep hole with high mountains all round it. Here we had a hot meal preparatory to tackling Red Mountain Pass. About 12:30 we started up the road which climbs abruptly out of Ouray on a series of switchbacks one after the other. It was now snowing hard, the wind was blowing, the temperature hovered well below zero. All told it was a hell of a night. The Boss had been raised in this and it didn't occur to us to ask him if we were being wise. We simply did what he thought best. In numbers of miles this is only a short trip but it made up in quality what it might lack in quantity.

The switchbacks didn't look bad in that swirling nothingness of snow, and we couldn't see what lay below the road on the side without a cliff, and it is just as well for had we been able to see what we were nearly skidding into every little minute, I'm sure we would all have died then and there of fright.

We finally arrived in Silverton a few hours before daylight. The Boss took us to the hotel and registered, saying we would have to stay here a night while we gathered up more supplies and made arrangements for a string of pack mules to haul supplies in to the mine.

We might have had some warning that first morning in Silverton, but it all went right over our heads. When the waitress in the cafe asked us if we were going to be around awhile, we replied that we were headed for the Highland Mary Mine to work. She looked at the cook who just shrugged. Then she said, "You boys miners?" as she brought us bacon and eggs. I said, "Well no, but we are going to learn the job." The Boss came in then and introduced us to both the cook and waitress and asked the cook where the livery stable owner might be. The cook said he probably had gone to the post office and if so would stop for coffee on the way back. It seems that the livery stable owner's business was the operation of several strings of pack mules hauling supplies to, and from, the less acces-

The Trilby Mine, high and lonesome, 1980. Zeke Zanoni Photo

sible mines in the area. In a few minutes, the man arrived. He too greeted the Boss as an old friend and asked him, "What the hell are you doing up here this time of year?"

"We're going to work on the Trilby Tunnel at the Highland Mary," was the answer, and "how are the chances at making arrangements for some of your mules to get our stuff up there?" The livery man looked blank for a minute, then said, "What do you mean, supplies up there? You know as well as I do that getting mules over that trail has been out of the question for over a month, and will be out of the question until probably next June!"

The Boss looked unhappy and admitted that he had known that would probably be the answer, but was hoping the snow hadn't closed things up yet. The livery man looked as though he was about to say more, but didn't. The Boss stirred his coffee silently while we sat there stunned. Our new jobs were going up in smoke in much less time than it takes in the telling. Then the Boss said, "There *is* a way, of course," and we waited for him to go on. "We can pack the stuff up on our backs. It's only a mile from the mill up to the Trilby bunkhouse, and it won't be too bad." It didn't occur to any of us to ask why mules *couldn't* go in November where they *could* go in June, or what reason he might give for thinking we could do what mules couldn't. In short, any job was still a job, and we didn't hesitate. Fools do rush in.

Next morning, we headed out of town with the truck loaded completely to the gunwales with supplies; more food, fuse, blasting caps, drill steel, blacksmith coal on the top of the load hauled from Denver, and on top of it all our bedrolls and bags. It was a beautiful morning, clear and very cold, ten below zero. The road was well plowed for the first few miles, and fairly well cleared when we left the main road for a side road headed south into what the Boss named as Cunningham Gulch, a gorge carved out by glacial action, perhaps a thousand feet deep, U-shaped, with nearly vertical side walls. It's a quarter of a mile across from wall to wall, and several miles long, ending at the Highland Mary Mine and rising steeply from there to a tremendous alpine cirque a couple thousand feet above the bottom of the gulch and well above timber line. At the beginning of this slope upward there is an abandoned

This is the Trilby Mine as it looked around 1909. When Tom and his friends decided to move in for the winter of '31-32, half of the structure had been moved down hill and out of sight by an avalanche.

Photo courtesy of San Juan County Historical Society.

mill and above the mill a caretaker's house. From there to the Trilby Tunnel was "only a mile," but what a mile! This last distance to the mill was a matter of bucking snow so deep that time and again we were stopped cold only to back up and take a run to plow further on. There, at the end of the road, close to the mill, the Boss showed us the Trilby Tunnel. He pointed, straight up! There, hanging precariously to what appeared to be a sheer cliff a thousand feet above us was a building which looked to be hanging by its fingernails, terribly temporarily.

I asked the Boss, "How long has *that* been there?"

Lightly enough, he answered, "Nearly forty years, or *half* the building, at least."

"Half of it? What happened to the other half?"

"I'll tell you all about it when we get up there where we can see it better." By this time you must be certain we were all insane, but remember, hunger and the need of a job do strange things to people.

Then came the caretaker to see what was going on. He greeted the Boss sourly, seeming to feel that he had been invaded. The Boss greeted him cheerfully enough, and introduced us to the man whose name was Palmer. Palmer had lived too long by himself, a watchman of closed mines accustomed to seeing no one for months on end. The Boss went on to say we would have to bunk with Palmer at his place for a few days while we were getting our stuff hauled up the hill and getting the bunkhouse ready for our use. Palmer looked dubious and the Boss said, "Haven't you had word from Craig?"

Palmer said, "Yes, I have, but I told him this whole thing is impossible this time of year and that he could forget about the whole crazy idea until spring, and he hasn't answered yet."

"We're your answer to that, good friend, so just pitch in and give us a hand getting the grub that will freeze and the bedrolls up there to your place."

Palmer, muttering to himself, did as he was told and by the afternoon we were installed for the night. Palmer had plenty of room as the building had originally been the mill bunkhouse. There were many rooms, but most of them were empty as Palmer had taken four at one end of the building for his use and closed off the rest to cut down on the heating. Cold as it was, we were cozy enough with a big kitchen range in two of the rooms, with coal in the bin to fire them. Palmer set to and cooked dinner for us and we all were more than willing to turn in early looking forward to the first trip up the hill in the morning.

Palmer had plenty of beds made up, so we didn't use our own bedrolls. This we regretted very quickly. Once the bed warmed up we found we were anything but alone, and learned what it's like to have myriad fleas hopefully eating their way along our bodies leaving trails of bites that reminded us for days of their presence.

It was cold when we rolled out of bed and got fires started, and black as black outside. Down here in the bottom of the gulch the sun didn't look in until almost noon at this time of year, and didn't stay long. Daylight came as we were finishing breakfast. The Boss decided we would take nothing along until we had made a trip to see how the land lay, in what shape the bunkhouse might be, etc.

So we set out on foot in snow too many feet deep. We only went a mile, but we climbed a full 900 feet in that mile. Finally we got within sight of the bunkhouse. We were standing at a curve in the trail that looked backward to a nice, safe, wide trail, and forward to a thousand yards of very steep slope over which we would have to pass to reach the bunkhouse. The slope was narrow with a cliff over it on the left, and a thousand feet of air beneath it on the right. There was no sign of a trail in the expanse of snow-covered rock.

"Here," the Boss said, "we will spread out. I'll go first and you all wait until I get a hundred yards out before the next one starts, and let him get the same distance out before the next, and so on."

Sam, one of us who was never much for talk, said, "Why?"

"So if the snow comes loose it won't get more than one of us."

"If the snow comes loose," he had said. Sam looked at me and I looked at Sam, and we both looked down into the chasm below the trail. It was clear what would happen if that snow comes loose! At this point several other things cleared up, too. Why, for instance, mules couldn't haul our stuff in, but *we* could. Mules cost money, but fools like us could be hired for a dollar a day.

Well, I was tired of going without a job, so when the Boss got out there a hundred yards I followed him, and when I got out a hundred yards, Sam followed me, and when Sam got out a hundred yards, John followed him.

After the first few hundred feet the sheer effort to stay on the trail overcame the freezing fear and by the time we got to the bunkhouse the fear was, almost, forgotten. It was high there, and we were all fighting for breath and sweating even though the temperature was still way down. The Boss dug the snow out from around the door of the blacksmith shop and we went in, got a fire going in the forge, then took a look around. The building was two stories tall, anchored to the cliff by cables bolted to pieces of drill steel, which in turn had been cemented into holes drilled in the rock. The building once was twice as long and we asked the Boss what had happened to the missing half.

"Snowslide chopped it in two."

"Any men in it?"

"No, it was closed down at the time."

The second floor was the living floor with a big room for bunks, dining room, and beyond that, the kitchen. The whole place was a mess. The windows were not broken out, but water had dripped through the roof and left great spreading masses of ice on the floor. The tarpaper sheathing inside on the walls looked to be completely soaked and then frozen, and we saw after we got a good fire going in the stove in the kitchen and melting began, that this was indeed the case. The Boss rubbed his hands together and said, "Right now this looks pretty hopeless, but you'll be surprised how much better it will be when we get a little work done."

Nobody said a word. So, back down the hill we went, loaded packsacks with carpenter tools, nails, and each of us took a roll of roofing paper on a shoulder and off we went. After a few days and many trips up that damn hill and across that last bit of treacherous trail we were beginning to get the place almost livable, so we began hauling supplies up the hill.

Another week of this and we got over the sore muscles that had gone with it. Nearly every night it would snow a foot or so, and most days it snowed part of the time so snow was steadily getting deeper and the trip up the hill was each time more difficult. As time went on we dwelt less and less on that chance of the snow going out over the dangerous section, but were always subconsciously aware of the simple fact that with each foot of snow laid down on that steep slope, the danger of its sliding grew more acute.

We were now hauling dynamite. We had a ton of the stuff to move, and taking the sticks out of the boxes and loading them into packsacks was a time-consuming chore. It was Palmer who suggested that he load the extra packsacks while we were on the trail, so we wouldn't lose that time each trip. None of us gave a thought to this offer of help from Palmer, who generally was grouchy and not too inclined to help. This was a good idea, so now we had only to shed the empty packsack, shrug into one that Palmer had loaded, and start back up the hill.

Dynamite is safe enough to handle. Nowadays, it isn't as ticklish as it once was. Time was a few years back, when dynamite would freeze and become mighty unstable stuff, liable to detonate for little or no reason, but not this modern stuff we were packing.

The trail became more and more of a problem. Have you ever walked on a trail built up little by little {from} snow falling day by day? The trail turns into a narrow track of solid material with nothing but fluff on both sides. If your foot doesn't land just right on the packed trail, it slips off into the fluff and you are suddenly waist deep, or chin deep, in soft snow, and with a packsack of dynamite on your back. Other times the foot slips into thin air and you land on the back of your neck—still with that dynamite, only this time it's underneath you, a cushion. It's a comforting thought that today's dynamite is really nice safe stuff.

Once across the treacherous last bit of trail, we removed the powder from the backpack and went back down the hill for another load. The third day of this went on fine until I casually dumped a load of dynamite out on the floor of the blacksmith shop and started to stack it into a box. Whoa up, here! What the hell? These sticks were discolored and gooey, what goes on?

Nice fresh dynamite doesn't look like this. I picked up a stick and turned it to read the date printed on its side. January 6, 1918! Where did this come from? Wait, fifteen years? My God, that goes back far enough to predate non-freezing powder, and if this has frozen and thawed repeatedly for fifteen years it is certainly terribly sensitive to shock.

Slowly and carefully, I laid the stick down and backed away from the pile, wondering where the stuff might have come from, and what to do next. The Boss came in and stood there puffing with his load of more of the nasty stuff. He became aware of my tension and said, "What's the matter with you?" I pointed to the pile of dynamite sticks on the floor. He had started to remove his backpack, but paused while he took in the significance of the discolored mess on the floor. Then he paled, turned round, went outside, walked to the end of the trestle from which rock was dumped into Cunningham Gulch and there, slowly, cautiously, he shrugged the straps off to let the packsack slide from his back down through the ties under the rails of the trestle, and watched it lazily turn over and over as it dropped a thousand feet to the rocks below,.

This is the high trail, we think, which wound around from the Old Shenandoah-Dives Mine above Cunningham Gulch past the Trilby toward Spencer Basin and the Highland Mary Mine. A nice trail in summer, 1904. Tom Savich Collection

There was a flash of light, and an instant later the roar of the explosion floated up to us.

The Boss turned to me and said, "Palmer! Palmer has traded some of his old powder for some of our new!" He started down the hill with the idea in mind of teaching the old man some sort of lesson, and the closer he got to Palmer's house the madder he got until he was ready to explode long before he reached the last bit of trail. When he saw the door standing open he deduced that Palmer had heard the noise and had abruptly taken off for parts unknown.

Now that it was safely over it began to have its funny side even as we had pictures of ourselves being blown to bits.

Hauling finally was finished. We figured we had enough grub to last us for two months, and if the weather permitted we planned to get in enough for a third month, which would have taken us close to the first of April if worst came to worst. The weather turned bad though, and we went to work in the mine instead of bucking the snow. So, we never got around to adding to our food supplies.

Getting into the mine tunnel proved to be another sizeable job. The tunnel had not been worked for many years so the rails and the mine cars were not in the best of shape. None of the wheels on the cars would turn, they were rusted solid and it took days of patient pounding to get that repaired. Then with tools loaded into a car, we pushed into the tunnel. We made perhaps a hundred feet, when we came to a pile of rock that had caved out of the "back"—the ceiling of the tunnel.

Water draining along the tunnel had backed up a foot deep at that rock dam and perhaps 200 feet into the tunnel. It was a beautiful blue, clear as crystal. We could see the rails submerged in it and they looked to be perfectly preserved. It took a day to move out the fallen rock and to drain off the backed-up water, but finally we were able to move once more. We made almost ten feet this time. When the car got to where the rails had been underwater all those years, the rails proved to be nothing but well-formed mush. The mine water, which contained a lot of copper salts, had replaced the iron in the rails leaving behind a copper compound that was nothing but a wet powder. Farther into the tunnel, there was a side drift that headed steeply uphill so the rails were high and dry.

We re-laid rails from that drift in the main tunnel, and so, finally after many days of work, we got back into the part where the Boss wanted to start actual mining.

Gold, silver, copper, lead and zinc occur in and are part of some rock. In order to get to where these minerals occur, one must drive a tunnel through solid rock. This is done by drilling holes in a particular pattern in the rock, filling the holes with dynamite, by means of which the rock is torn away, broken into pieces. This rock is loaded into mine cars, hauled out and dumped. A new set of holes is then drilled and the process is repeated. Each set of holes drilled by hand advances the tunnel perhaps two feet. Eventually, one hits ore, which is rock rich enough in desirable minerals to be mined at a profit, and tunneling then goes on in this ore. This is oversimplified, but may give an inkling of how this is done.

Today, in mines everywhere, machinery is used which makes this whole process relatively simple, and fast. A machine drills the holes, another machine cleans up the muck and still another machine hauls a train of cars away for disposal. Our methods were much more primitive. First, holes had to be drilled using a four-pound hammer propelled by elbow grease. It sounds simple enough? Well. We beginners moved into this thinking much the same, but we learned in the first fifteen minutes that we had a man-sized job on our hands.

Holding a drill in one hand, hitting the drill with a hammer held in the other hand, doesn't sound difficult, but, oh my. It doesn't take too many strokes before the weight of the hammer grows until you are moving a mountain each time the hammer is lifted. When the mountain gets big enough, one's aim tends to go bad, and then the hand holding the drill receives the blows intended for the drill. Soon the holding hand is a bloody, bruised mass of bone and muscle.

This is true only until you begin to drill "uppers." These holes "look up" as the miners say. Such holes are drilled by holding the drill up against the rock with one hand and striking the drill with the hammer swung underhand. Picture this. Some fun! This sounds easy enough, but how hard it is to hit the drill! Somehow, the holding hand absorbs most of the blows, and soon both sides of the holding hand are bruised and bloodied. When you think this has gotten as bad as it can get, the skin starts to come off the inside of the holding hand, and when that is good and raw the blisters on the hammer hand begin to break.

I must advise that the rock was so hard the steel seemed to bounce back at you rather than to cut rock. Sam, tougher than I, managed to drill five inches of hole in the first ten hours of effort, while I could claim only four! I'm left-handed and Sam is normal, so we worked side by side in the drift, I on the left, he on the right. Ten hours of swinging a single-jack (seems) not less than four million years, and that's no exaggeration. I thought I had seen some tough work before I hit the mine, but after a couple days of single-jacking I decided that all the other work had been child's play.

At the end of that first long, long day we walked out of the tunnel with hands held carefully well away from our bodies. Washing up was done, but only just enough to see that thorough washing was out of the question. Fortunately we had beans for supper and could pour them into our mouths directly from the bowl, holding the bowl with the wrists. Hold a fork? No way! By the time we had been out of the mine an hour we couldn't even move our fingers let alone pick up a tool. The Boss took pity on us and filled a couple of bowls with ice water, dumped in plenty of salt, and made us soak the hands. What we called him is unprintable, and we were sure he was ruining us for good.

By morning we decided the treatment had helped a little. At long last morning did arrive and we seemed hours getting into our clothes and breakfast eaten. We delayed as long as we could but finally there was nothing left to do but go back into the mine and to those hammers.

The second day was, if anything, worse than the first. Now, in addition to battered hands we had endless muscles announcing their misery. By noon, we were sure there was no earthly way of surviving to the end of the shift, but live we did, and we drilled and inch or two more than on the first day. The Boss was disgusted and said that he might live long enough to see us qualify as miners, but sincerely doubted it. We didn't give a damn what he thought, but were too tired to say so. This night was worse than even the day had been because we were now so sore in so many places that sleep seemed out of the question.

In spite of that, day came much too soon, and in we went for the third day of torture. This day too was endless, but not too horrible, and a couple more days found us finally glad we were alive and looking forward to the time when we would complete the first "round" of holes and get them shot. It seemed we would never get all those holes drilled, but finally it was done.

The holes were loaded with dynamite, fuses carefully cut to length so the holes would shoot in the right order, the fuses lit. We scurried down the tunnel far enough to put a couple of bends between us and the "face" where the explosions would happen. We stood and counted the blasts to make sure that if there should be a "miss" we would know about it when we went back in. Otherwise one of us might hit the un-detonated cap with a pick or shovel and have a quick ride to Kingdom Come. The count was right, though, and we walked out of the mine feeling we were beginning to accomplish at least a little.

The next morning we went in and at once ran into the smell of smoke left from the night's shots. In a large mine there would have been forced ventilation which would remove the smoke during the off-shift, but here we had to depend on natural circulation to do the job. At some time in the past a shaft had been driven to the surface from well back in the mine to allow mine air to move up and out of the mine. Cave-ins had nearly closed the passages leading to this air shaft, so circulation was limited. It was enough to remove the worst of the smoke and its carbon monoxide content and we got in to the work place without trouble, looking forward to seeing an enormous pile of shattered rock.

The pile wasn't much, and as we approached the end of the drift we saw that our holes hadn't done too well. Instead of breaking rock cleanly to the bottom of the holes, the explosion broke only the outer few inches of the hole, leaving the remainder relatively untouched. These are called "blowouts" and result from wrongly placed holes, or under-loaded holes, or both.

Net result: Most of our horrible first days of drilling had been wasted! We mucked out the broken rock, and started again. The second round was better than the first, and before long we were getting good results most of the time.

Sensing some progress made, the Boss decided it was high time to do something about the decreasing supply of grub. He was our blacksmith and although he had been training John, we had our misgivings about his leaving, but food was needed, and if the Boss went, we could go ahead with the work. He figured to be back in a couple of days, at most. So, in the face of ever-deepening snow, he took off. We watched him cross the dangerous stretch, or almost cross. Falling snow hid him completely before he had gone far. We heard no rumble of snow moving and figured he got safely across.

As feared, we began to have trouble with the drills, "the steel," so-called. We organized the work with Sam and I working in the mine while John did the steel sharpening and tempering, along with the cooking. Steel sharpening isn't too difficult but tempering requires a sharp eye and a deft hand. Without these, drills can be bad. One drill would be so soft it would rivet over like putty, and the next so hard and brittle it would shatter at the first blow.

The main trouble with John was that he knew everything. You know? One of those. We tried to tell him what was wrong with the steel, but he could never admit that *his* steel could ever be anything less than perfect. The fault had to be ours, we just didn't know how to handle good steel! The time came when John decided he would have to come into the mine and show us how this thing should be done. In he came, and stood around watching us work. This proved to be the rare day when the steel was just right and soon John began to crow. Sam listened as long as he could, trying to think of some way to get John out of there short of killing him outright. Sam found the way, so simple it was perfect. He said, "Look here, John, there's no use in your standing there doing nothing. You hold steel and we'll get busy with the double-jacks." A double-jack is a seven-pound hammer with a long handle. A man uses both hands and swings the hammer up over his head. Sam and I had been at this long enough to be hardened up, but John had not. We poured it on and although all he had to do was turn the steel after each strike, that was more than enough. He was game but in an hour his hands were raw and bleeding and we gave him no quarter. We kept him hard at it until the end of the day.

John was glad to let me fix the food and he was off early to bed in total silence. The next morning he got breakfast and was out of sight when we went into the mine. Steel was much improved from that day on and John never went underground again.

Life in the bunkhouse was not too complicated. Our beds were boards ripped from the north end of the bunkhouse, no springs. Nights went far below zero and the temperature inside went right with it. Once we had settled down, the pack rats started in and it sounded as though they were playing football over and around our beds. Once the stove cooled, they climbed in it hunting food and made a great racket. Sam had a twenty-two pistol and he went to bed with the gun and a flashlight alongside his bed. When none of us could sleep for the noise, one of us held the light while another shot at rats. It was a little hard on the stove, but it wasn't in too good shape to begin with. It was one of those old wood burners with a big door on the front where big chunks of wood could be loaded into it. This door had a makeshift replacement for the handle long gone, a white porcelain insulator. One of these nights we heard a rat playing around on the stove and when we got him in the light he was perched on the insulator.

Inside the Trilby Mine in 1991. On the left is an ore chute that handled rock from the stope above. The broken ladder and two inch air line once served men working two hundred feet higher. *Zeke Zanoni Photo*

Sam shot twice, missed twice, and the rat just sat and ignored us. The third shot hit the insulator, knocking it into a million pieces and knocking the rat up onto the stove top. From there he jumped up in the warming oven. Sam tried a couple more shots but only perforated the back of the oven. The evening's shooting ended with my getting up and closing the oven door on the rat. We disposed of him in the morning with a club.

Mornings, the first fire was built by the loser at cribbage the night before. Crawling out of the mountain of bed clothes into the early morning chill was hard to do, so the cribbage games were played in grim earnest. Somehow it usually worked out that John had the fire building to do and we would lie in the sack and gloat while the poor guy froze. As winter set in earnest, the nights got colder and colder until little clothing was removed before we got into bed. You can imagine how fragrant that place got with the variety of smells that developed with the passage of time.

Snow in the air was the rule rather than the exception. It was nothing for it to snow a week at a time without letup. When it wasn't snowing the wind blew and moved the snow around until it seemed to be nearly always snowing. With the weather outside so lousy it was easy to concentrate on the work in the mine without paying a lot of attention to the condition of our grub supply. The Boss had not returned, but we didn't worry because we figured we could always get down the hill even if he might not be able to get up again.

We had learned from the Boss that Cunningham Gulch was avalanche prone, though during the dead of winter the big slides seldom ran. Smaller slides, usually resulting from cornices at clifftop built up by the wind until they couldn't support their own weight, were common this time of year. Above that last stretch of trail we watched a cornice building from day to day. Since these smaller slides usually ran during stormy, windy periods, it was unwritten law that no one moved outside at such times.

The wind did tricks with the snow. One morning we awoke to find the kitchen filled with snow, completely burying the shovel. The wind had sent a trickle of snow running down a seam in the cliff above the bunkhouses, down across the steep roof to the kitchen chimney. The chimney had half-inch holes in its rim to prevent the stovepipe from overheating and setting the roof on fire, and through these holes the snow trickled through the night, filling the kitchen to the ceiling. Breakfast was a little delayed that morning.

The Boss had gone down the hill about a week before the stove got buried, and hadn't returned. We had no way of knowing what had happened. Maybe he couldn't get back, but in any case we were running out of grub. The snow continued, the cornice above the trail grew ever more ponderous. We were caught between a rule that said we shouldn't attempt to travel, and a lack of food. Finally everything was gone but soda crackers. The pack rats had been at these, and in fact had packed much of the cutlery from the kitchen and left it in trade for crackers taken and eaten. We brushed off the remaining crackers and made do with them as long as they lasted.

We finally decided we would try to make it out. We spent a day and most of a night building coffin-shaped boxes and lining them with pieces of ten-inch ventilating pipe taken from inside the mine and flattened out. Our belongings went into the boxes, hoping the lining would keep the rats and moisture out until we returned in the spring.

Bundled to the ears with everything warm we owned, we set out in the teeth of a blinding snow being tossed wildly around by the wind. We figured that if we could get safely past that cornice we would have easy going the rest of the way.

When snow is deep and new snow forms new layers on top of the old, the snow settles. Not steadily, quietly, but suddenly in abrupt jerks. It will suddenly drop a few inches underfoot with a loud "whoof." When this happens it isn't a small area of snow, but usually an area large enough that when it happens it feels as though the bottom has dropped out from under the feet. On a steep slope where this kind of settlement can trigger sliding snow, it is a very frightening feeling and sound. By tacit consent we didn't bother to separate on the trail. Somehow none of had any desire to be separated from the others. We weren't twenty yards onto the trail when the first bit settled under us, and we all froze, wondering what would happen next. Nothing happened, and we moved on.

Sam was in the lead, I next, John bringing up the rear. As we moved under the overhanging cornice, a slab settled under me and began to move down the slope. I thought, "Oh–oh, here we go." I looked up to see Sam turning away from me and thought, poor Sam, he doesn't want to watch me go over the edge. The motion was slow but inexorable. I was waist deep in snow and there was no question of being able to move off of the shifting slab. "Nothing to do but pray," I thought. When I was perhaps fifteen feet below the trail moving slowly and slowly gaining speed, my feet caught on a rocky rib sticking up into the snow. There was no time to do anything. It all just happened. Instead of sweeping me off my precarious perch, the slab split in two and passed around me. The mass of moving snow went over the edge below me with a roar. With a sigh of relief I looked up to see if that cornice was still hanging there and Sam hollered, "Let's get the hell out of here!"

I scrambled to get back up to the trail, we all scrambled to get out from under the cornice. When we were almost in the clear, it let go and dropped onto the slope taking everything with it. We kept going, trying to hurry and to look over our shoulders at the same time. There was no way this was going to miss us, no way. This was so clear, so certain that we just stopped and faced the monster bearing down on us. It was quickly all over with. As the enormous mass went over the edge we found ourselves standing on a little island around which the slide had veered, leaving us white and shaking, having had a very close look at the old man with a scythe.

"Well," John offered, "at least it won't be as bad from here on down." We moved on down the trail, floundering in very deep powder snow. A foot placed off of the trail now left us down over our heads in snow, and fighting up out of that took enormous amounts of energy. We had only three-quarters of a mile to go, but it went terribly slowly, steadily draining our energy. We sweated as though it was ninety degrees instead of fifteen below zero. I traded with Sam to break trail and to discover how nearly impossible it was to move. Soon we all tired, John was first to sit down and just quit. We fought him onto his feet, made him keep moving. Time and again he went down, and each time it was harder to get him back on his feet. When it seemed we could go no further, Palmer's house came in sight. Certain we could get that far, we all sat down in the deep, deep snow and rested.

At Palmer's we got a fire going and as the place warmed we shed clothing. Sam said, "Do you realize that mile took us four and a half hours?" It didn't seem possible, but he was right. We dug into the food cache, ate pork and beans with great relish, and dropped gratefully into bed without a thought for fleas, thankful only for having been allowed to get this far.

Next morning we held a council. It was four miles to the nearest place we could expect to find people. The snow had let up a little, but now the wind was really howling up the gulch and this meant that almost certainly slides would be running, dropping into the bottom of the gulch where we had to travel. It was a risk we had to take. The food cache wouldn't keep us more than a day or two and there was no real hope of major change in conditions in that short a period.

There were several pairs of skis in the cabin, but no snowshoes. We worked to rig bindings and with this little preparation, set out. Skiing under these conditions was no sport. The skis supported us a little, but we travelled in snow to our armpits most of the way. That meant pushing an enormous load of snow ahead of each leg as it pushed forward and this used muscles unaccustomed to such treatment. Before we reached safety the large muscles on the front of our thighs were rolling up in agonizing cramps with every step we took. We traded around on the trail-breaking, but now there wasn't a lot of difference between leading and following.

This four miles took us nearly five hours. When we finally arrived at the Old Hundred mine we were on our last legs. The folks there, a watchman and his wife, met us with open arms, gave us coffee laced with bourbon, and called Silverton for a car to come for us. They made sure our feet were not frozen and kept the coffee and bourbon coming.

My memory from this point on is hazy (it couldn't possibly have been the bourbon) but I came to in the hospital in Silverton. Sam and John were in beds alongside. We were all OK, just needing to recover from exhaustion. We stayed one day in the hospital, then moved over to the hotel for two days of rest.

Just being safely in Silverton again provided no relief for Tom Trumbull and his companions. The second part of the article went on to tell of the trials and tribulations the men faced upon trying to ride the train the 45 miles to Durango, with the hopes of eventually getting back on home to Denver. The winter of 1931–1932 was an extremely difficult one, even for a town as tough as Silverton. The journey proved rife with starts that were stopped by avalanches that forced returns to town. The journey, made today in four hours by train, took five days to complete.

The gentlemen, finally did reach home. They actually turned around in the springtime and came back to Silverton to finish what they had started until the mine's backing disappeared and once more they were forced to leave town.

Tom Trumbull went on to graduate from the Colorado School of Mines with a degree in mine engineering. He also was a rancher. In 1938 he married "Bobbie" Felder. Tom moved to Grand Junction in 1991 where he passed away on May 30, 1994, at the age of 88. ❋

Tom Trumbull, 1994. Photo and story courtesy of Tom Lindquist

(left) Granny lived in this cabin when it was in good repair in the late thirties, early forties. Across from her was the Hamlet Mill. The road ran in front of the cabin and the railroad ran behind. Photo from Jody Sutton, Wilma Bingel Collection

Wilma Bingel's mom, Rose Etta, known to all as Granny Grey, was full of energy. She loved to fish and every Sunday she would climb the steep trail up to Hematite Lake. Her last trip was in 1971 at the age of 71! Have you tried that hike? Wilma Bingel Photo

Well, glad *we* were up Cunningham in the summer. Getting down will be a little easier. Back from the Trilby, down past the Highland Mary down the valley and there's the train waiting at the Green Mountain Mill. Now, we'll back down towards Howardsville, get on the main line, go past the depot and head north to Eureka. Yes, depending on the years, there were buildings and mills and little towns along the three mile stretch between Howardsville and Eureka.

EUREKA

Up the road by Maggie Gulch and before Kittimac was another cabin. This is probably the Toy Family around 1921. John McNamara Collection

Helen Bell Watson was the sister of Jim Bell and a granddaughter of Ruben McNutt, Silverton pioneer. *Gerald Swanson Collection, 1925*

Not a good day to hang the wash out as the Silverton Northern comes steaming and smoking into Eureka, a town apparently named by Ruben McNutt, Jim Bell's grandfather. Look how little snow there is. Could this be June? Hmm, maybe not. Anyway, those are company-built, family homes shown here. Tract housing, sort of.

Silverton Northern Train Photo, San Juan County Historical Society

Now we pass Hamlet, a few cabins, the Hamlet Mill, Middleton and Maggie Gulch, the Kittimac Mill, Minnie Gulch—Maggie then Minnie—Why is that confusing? It's alphabetical. Maggie is closest to Silverton. Here comes Eureka!

'Eureka!' exclaimed Archimedes, on discovering a way to determine the purity of gold. Today's interpretation of the word is generally agreed upon to be *'I have found it!'*

Jim Bell does a little prospecting, probably around Beartown and Kite Lake. Like his grandfather, Jim's interests were in mining and milling and assaying. Ruben died in 1909, Jim in 1988.

Steve Meyers Photo, Marge Bell Collection

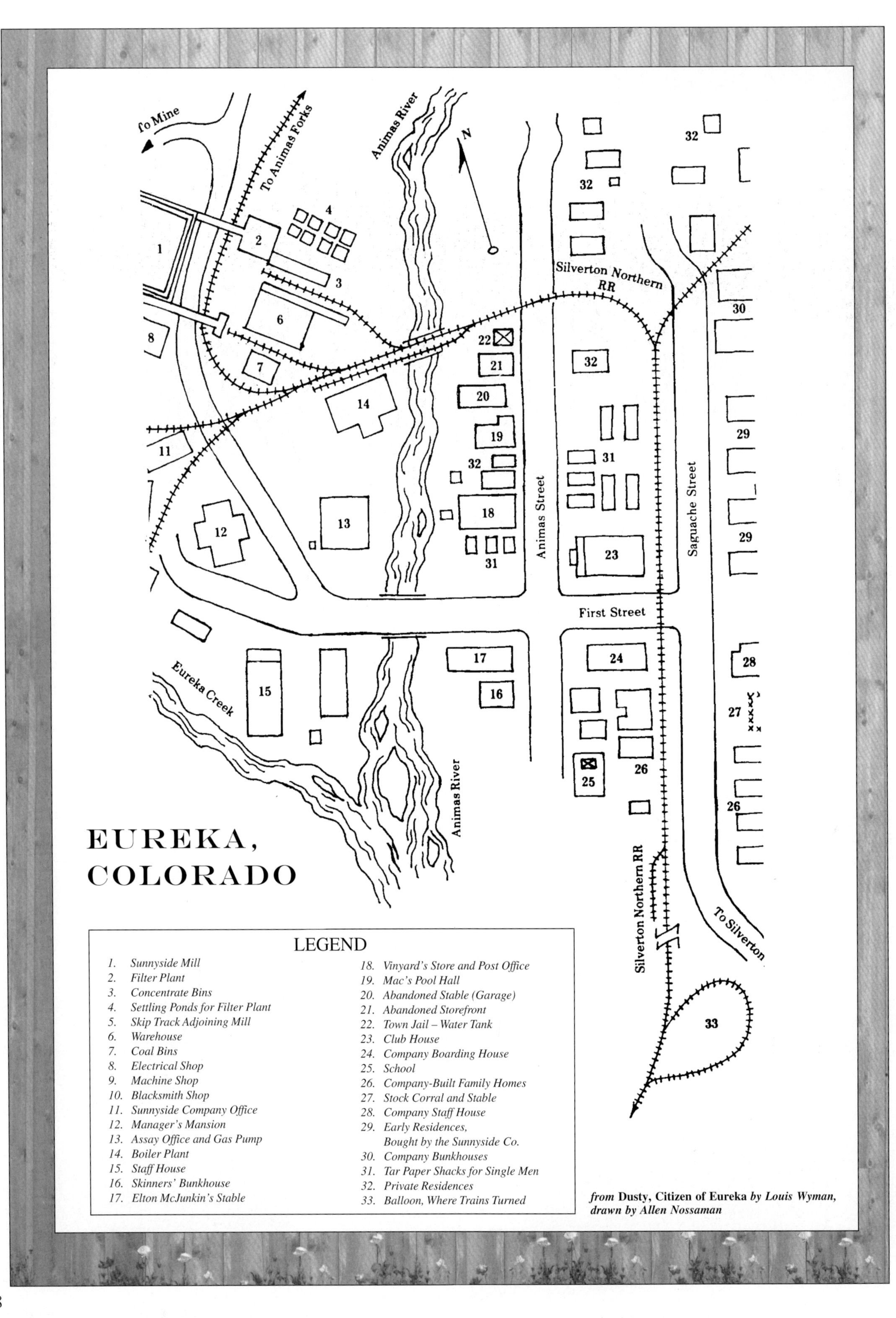

from **Dusty, Citizen of Eureka** *by Louis Wyman, drawn by Allen Nossaman*

EUREKA OWED ITS EXISTENCE TO THE SUNNYSIDE MINE. Like mind and body, they are inseparable. Even though the Sunnyside Mine was one of the longest operating, most heavily producing mines of its kind, its classical, chronological history dispels the presumption that owning a mine was synonymous with getting rich quick and owning a fortune:

1873–While still Indian Territory, two men, McNutt and Howard stake a claim to the Sunnyside Mine, located twelve miles northwest of Silverton.

1874–Howard sells out to McNutt. McNutt grubstakes (in exchange for food and mining supplies for any future profit) with Engleman, the sole merchant of nearest town, Eureka, seven miles north of Silverton.

1875–Greene Smelter, San Juan's first smelter, opens in Silverton. Sunnyside can now process its low grade ore of silver and lead nearby for a profit.

1878–Silverton's poor grade coal makes Greene Smelter inefficient. As with any smelter, mines receive only one-third of smelter's initial quote. So far annual work required to maintain claim to Sunnyside ($100 dollars per year for five years) exceeds value of ores produced.

1879–Engleman buys out McNutt then sells one-third interest to Thompson to work the mine. Thompson immediately sells all but one-twelfth of his one-third interest to Eastern capitalists to raise money to work the mine. It takes two men ten hours a day to advance a five foot by seven foot tunnel six inches. With advice of Porter, a well known mining engineer, it's decided to move Greene Smelter to town of Durango (which didn't exist yet) to start operating smelter when Denver Rio Grande Railroad arrives.

1881–Denver & Rio Grande Railroad arrives in Durango; Durango smelter starts up. Ores arrive by mule train.

1882–Denver & Rio Grande Railroad arrives in Silverton—can now ship by rail low grade ores to more efficient Durango smelter using Durango's high grade coal (Bodo Park).

1883–Just as things are looking good, metal prices plummet to a penny a pound for lead—no profit possible. (170 feet of mine diggings yields 500 tons of accumulated "dump" to be packed by burro down to Silverton for train delivery to Durango smelter. It's decided to pound or process the ore on site with a stamp mill and extract gold on site. Each "stamp" is a gravity-driven jack hammer, caused to lift and fall onto the rock, crushing it into sand. Mine operator requests Joe Terry, an expert mill man to build a mill at the mine site.)

1888–Thompson's mill completed at Sunnyside. Coal is to be the energy source for steam driven ten stamp mill. Coal shipped in by rail, then packed in on the backs of burros. Coal is expensive but is a year-round source of fuel.

1889–New mill crushes eighteen to twenty-four tons a day. Joe Terry foresees the cost of coal breaking the camel's back and builds his own Midway Mill (halfway down the mountain between Sunnyside and Eureka). His stamps run by hydroelectric power of Eureka Gulch which costs nothing to run but is a seasonally dependent water flow. Twenty-two miners create fifteen tons of dump a day. This dump is transported from Sunnyside to Midway Mill by twenty-six burros making five trips a day. An entire day's fifteen tons yields 2.5 ounces of gold—considered good yield. Midway Mill runs three months a year.

1890–Thompson, Sunnyside's operator/minority owner is going bankrupt. Twenty-two miners owed months of back pay. Thompson owes money to every bank and store in Silverton. He mortgages potential silver, lead and gold to be extracted from accumulation of winter's dump. Ends up one-third short of paying off mortgages. Thompson is in desperate financial straits, with his health failing.

1890–Thompson dies. Terry buys Thompson's 7/36 interest in Sunnyside and assumes operation.

1892–Terry about to lose Sunnyside to foreclosure and back taxes. In last ditch attempt, Terry mortgages his Midway Mill for $3,200 dollars and has six months to pay. He uses $2,900 dollars to buy back 7/36 interest in Sunnyside.

This is a level of the 4th of July stope near where a rich vein of gold was found in 1901. That vein kept the mine open and running for a few more years. *From Allan Bird's book,* **Silverton Gold.**

1893–Government cancels Sherman Purchase Act—government will no longer purchase silver. The Silver Panic of 1893 results, prices of silver plummet, price of lead falls to 45-year low, production of all mines slows to a trickle, many railroads go bankrupt, the whole nation suffers one of the worst depressions.

1894–Terry convinces creditors to hold out for future profits, otherwise their investments are a total loss.

1895–Terry's relentless tunneling uncovers a large body of native gold; the mill mortgage is paid off, creditors are paid off. From here on out, Terry's goal is to buy out all other investor's interests and sell the mine to Eastern capitalists. As long as the gold pocket holds out, Terry buys out other owner/investors. He now owns 22/36 of the mine.

1896–The Silverton Northern reaches Eureka.

1897–Terry sells the mine for a $100,000 dollar option, but the sale falls through. Terry pours $100,000 dollars into further development: ads stamps to Midway Mill to increase capacity to fifty tons per day and builds an aerial tramway from Sunnyside to Midway Mill to transport ore. No more burros. The tramway revolves and powers itself by the gravity pull of loaded buckets, the only cost is maintenance. Snow avalanches take out towers regularly though.

1899–Terry builds a new additional mill at Eureka, 125 tons per day, powered by water from Eureka Gulch. He extends the aerial tramway to run three miles from Eureka Mill to Sunnyside Mine, costing $75,000 dollars. The mill is ready to roll, Durango Smelter closes for strike and puts 650 miners out of work. In October, when the strike ends, drought shuts down Eureka Mill due to insufficient water from Eureka Gulch. Terry builds a new waterway of flumes from the Animas River 6,000 feet away. The mill shuts down again because the newfangled roller system is not working and is forced to replace them with old fashioned stamps. Shut downs are costly, $100,000 dollars is spent, and Terry is forced to again mortgage the mine and mill for $33,000 dollars.

1900–Electricity comes to the mines, outbuildings and mills.

1901–Terry, on the verge of financial collapse, strikes a rich vein of gold on 4th of July. He orders twenty additional stamps for Eureka Mill.

1902–Terry makes an all out push to sell Sunnyside. Terry's operation now consists of 16,000 foot tramway, two hydroelectric powered mills: the fifteen stamp Midway Mill and the forty stamp Eureka Mill, boarding houses for one hundred men, machine shops, telephones and electricity to all facilities.

1904–180 men employed, daily concentrate equals fifteen tons, highest grade ever known in the county, stamp mills run 365 days a year, twenty-four hours a day.

1907–Production at Sunnyside is waning, having dropped to seventh place, less than a carload a day of concentrate. Sunnyside now consists of almost ten miles of underground workings.

1910–Terry dies and ownership passes to his three children.

1917–The end of the family operation comes when United States Smelting and Refining Company buys the Sunnyside for $667,000; Terry's heirs receive mostly stock interest in the mine, not money.

1919–A devastating fire causes $400,000 in damages.

1930–U.S. Smelting closes mine and mill down.

1937–Mine and mill reopen for two years. Closes again in 1939 and is idle for twenty-one years.

1942–The Silverton Northern train tracks are removed.

1959–Standard Uranium (Metals) starts driving American Tunnel at Gladstone to get under old workings of Sunnyside.

1985–Standard Metals loses mine after twenty-six years of operation.

1991–Sunnyside Gold Corporation, the last corporate owner, closes the mine.

1873–1991 Sunnyside Mine sustained generations of families but no millionaires were ever produced. ❋ *—Zeke Zanoni*

Powder boxes from 1928 found on the #4 Level above Lake Emma. *Courtesy Allan Bird*

The above picture of Eureka was taken by the light of the moon, probably in the twenties, looking down from the upper level of the mill on the opposite side of the valley (see photo below). Silverton lies about nine miles to the south (towards the right in the above picture). Animas Forks is up the canyon about three miles to the north (on the right hand side of the picture below. *John McNamara Collection*

Mac stands in front of his pool hall. From the book **Dusty, Citizen of Eureka,** *by Louis Wyman.* John McNamara Collection

Johnny Mac, born in Eureka in 1917, pictured at one year of age. John McNamara Collection

FOR ME, EUREKA HAD COME TO MEAN JOHNNY MAC

Mac and Johnny Mac—the babies were always cuties and the men were often opportunists. The store below was purchased by Mac, Johnny Mac's grandfather, in the early 1900's from John Curry. Curry had started the first newspaper in Silverton in 1875 called* The La Plata Miner. *James B. McNamara, Sr., kept a small post office in the back and took any money the store might generate and invested it in claims up by Ophir Pass. When they lost more money than the store took in, he moved out to Eureka where Mac's Pool Hall became a local fixture until his death, in the store, in 1929.

A store of the early days, located at 10th & Greene, Silverton. John McNamara Collection

I had not been in the country when the town was booming, yet Johnny Mac had. I was lucky enough to get to know Johnny Mac better and better as time went along. The pictures and stories he so generously shared easily transported me to another time. Then all too soon he was gone. So this little section is primarily because of him and definitely for him. Thank you, Johnny Mac.

Johnny Mac grew up in Eureka. When he wasn't going to school, he often worked in his grandfather's pool hall. Mac's Pool Hall. The miners gathered at the Pool Hall before and after work. John, no doubt, picked up some ideas from them. Of course, this wasn't one of them.

BY THE RAILROAD TRACKS NEAR THE SCHOOL THERE WAS A SCRAP PILE OF FASCINATING JUNK—A POTENTIAL PILE OF MISCHIEF AND ADVENTURE, ESPECIALLY FOR JOHNNY MAC AND HIS GOOD BUDDY DUDE GALLAGHER. In this pile of metal parts they found old journals. (A journal is a part of a rotating shaft, axle, roll, or spindle, that turns in a bearing.) Johnny Mac and Dude fished a bunch of grease out of those journals. For war paint? No, they spread a nice coating on the tracks for they knew the train would be coming by pretty soon, building up steam for the steep haul up to Animas Forks. That is, until it hit that grease! Why, when the train hit the grease, the wheels would spin, and because it was a forced draft boiler, the fire in the boiler would blow out and the darn train would lose all its power. It takes a lot of time to build up steam all over again.

This time, John and Dude were sitting quietly, safely in school—they thought—when the door burst open and Pat Lonergan (son of Patrick) and Blackie Plantz (maybe Ralph Plantz, Blackie's father) barged in with dark scowls on their faces. They had just come from running the Silverton Northern and after speaking briefly with the teacher they turned and walked slowly down the aisles, closely inspecting each child's hands. John and Dude were totally unprepared for this and that black grease from the journals showed plainly under their fingernails. Caught! Guess you could say they were nailed. ❋

Johnny Mac (left) and Tom Kimball on an early bug made out of a Model T, 1924. **John McNamara Collection**

The following story from Charlie Thomas, who also spent his youth in Eureka and in Silverton, especially sticks in my mind.

I remember the first time I saw Johnny Mac. It was out in Eureka. He was riding his bicycle and had a big basket over the handle bars. He spotted three kids skinny dipping in the river. I watched as he snuck up on them, grabbed all their clothes, threw them into his bike basket and took off. That was Johnny Mac, all right. ❋

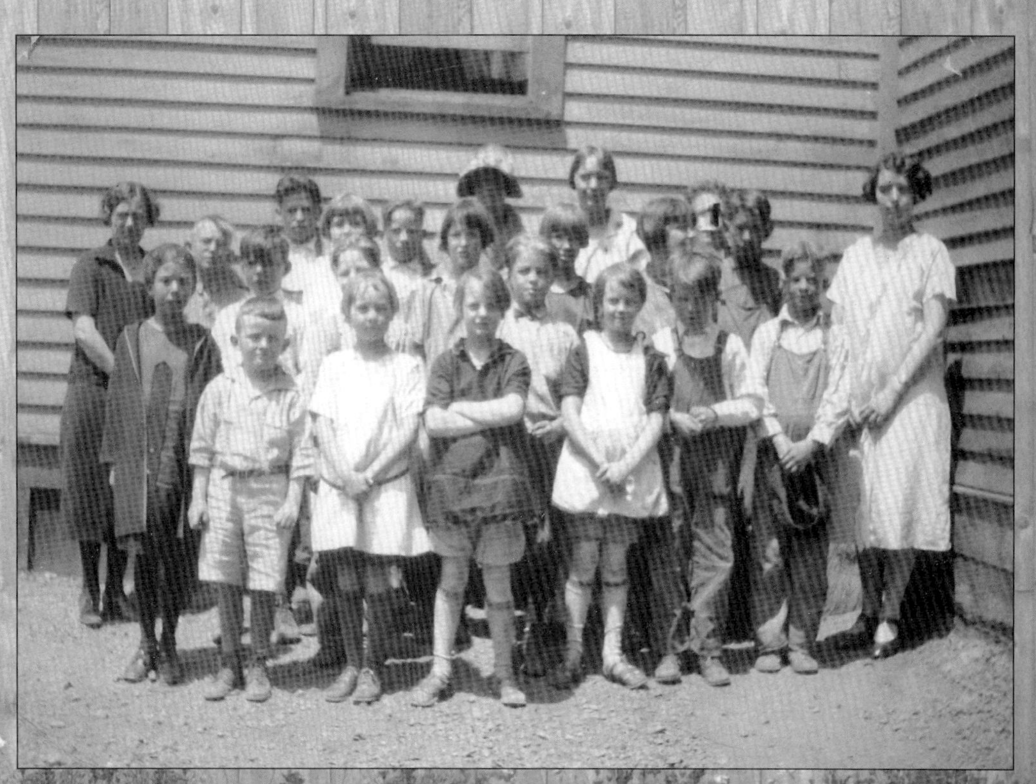

About twenty-one children attended the small, two room school house at Eureka around 1924. Grades one through four were on one side, five through eight on the other. In 1929 there was a ninth grade, then in 1930 the mine closed and so did the school. **John McNamara Collection**

Animas Street, downtown Eureka, a ghost town in 1935, about five years after the Sunnyside had closed. The mine reopened briefly in 1937 and 1938, but that would be the last time it was operated from Eureka. The next time Sunnyside operated it would be out of Gladstone.

Animas Forks is up the canyon to the right. The railroad crosses at the bottom of the picture. The water tower stood in the background (rebuilt today), then the jail and firehouse. The tracks were torn out in 1942. As we move from right to left, the next building was an abandoned storefront. The two-story building was first a dance hall and then a car park garage. Everybody in town who had cars could winter them here, up on blocks. The white building with two windows upstairs was Mac's Pool Hall, James B. McNamara, Sr., proprietor. The single story residence served as a shoe repair store belonging to Tommy Grenache. There was another single story building next to Tommy, purpose unknown. The last tall, two-story structure was known as Vinyard's Store and Post Office. The first building on the left, attached to the store, belonged to Bob Vinyard, post office and store operator (refer also to diagram page 118).

John McNamara Collection

John and Irma J. McNamara, after 54 years of marriage, 1994. When John first met her in Eureka she was a young girl on crutches. So he went into his grandfather's store and bought her an ice cream cone—they only had ice cream once a month in the store. She politely declined stating she didn't like ice cream. Years later when John moved to Durango he remembered that encounter with her. He looked her up and they were married in 1940. *John McNamara Collection*

Tough, funny, kind, talented—so many words fit him so well. Johnny Mac outside his Durango home, a couple of months before he died, December 11, 1994. *John Marshall Photo*

Daring riders on the Sunnyside tram, about 1920.
Tom Savich Collection

THE SUNNYSIDE TRAMS—COLORADO'S EARLIEST SKI LIFTS

The Sunnyside tram stretched 3° miles between the town of Eureka and the mine. Between the mine at Lake Emma and the long Terry span were two tension or weight stations. Their purpose was to keep the cables taut. The lower station still stands today. About halfway down the hill from the mine was an angle station. This was located just above Midway. Originally a mill was located at Midway (today the site is mostly boards and a foundation) and the tram ran from there down to Eureka. When the big mill was built in Eureka in 1918 the angle station was also built and became the main route between the mine and the Eureka mill.

The Sunnyside tram carried many buckets but their capacity was only 600 to 700 pounds. You could ride them up but you couldn't ride them down—unless you were dead or injured. The ride down was reserved for buckets of ore. In 1938 (when Sunnyside briefly reopened) you could make $4.40 a day working on the tram. For $1.50 a month, you could get hospital care for the whole family. Four men worked the tram at the mill end, four at Midway and six up at the mine.

Louis Wyman recounts the following story in his wonderful book about an adventurous dog, Dusty, Citizen of Eureka. *Louis knew the town and the mill and mine for he lived and worked there while Gustavson was the tram boss.*

"John Gustavson reminded me of a piece of whalebone—light, resilient and tough. (It was John who got Dusty to ride the tram.) The tram cables span the Eureka Creek Canyon at a dizzy height in their approach to the Midway angle station. The buckets go soaring along hundreds of feet above a rocky, timbered gorge far below. Until the last tower before the big span, the tram line maintains a fairly even height above the ground. Dusty (in his first ride) was becoming accustomed to it. But when he and John sailed out into space over the cliffs and timber, he actually cried, begging John to put him on the ground again. Dusty, the dog, often had the same feelings as humans. That particular feeling was one he shared with many humans. And that was just the angle station at Midway. The next stop was the mine. The man at midway pushed the carrier through the angle station and gripped it out sending John and Dusty on the last leg of their trip to the mine terminal. Sure enough, Dusty, certainly no wimp, had his nerve quit him again when the carrier soared, for the second time, some three hundred feet above the Terry Tunnel buildings. Soon however he was almost back to normal, giving each tram-tower a sharp bark to keep clear of their bucket. The brakeman at the mine tramhouse ran their carrier in on a side rail. And again, like many other tram riders, the pair headed to the boarding house where there were kitchens, cooks, and dining rooms."

Me? I'd rather take my chances with a tram any day than have to ride the quietest, emptiest freeway around. ❋

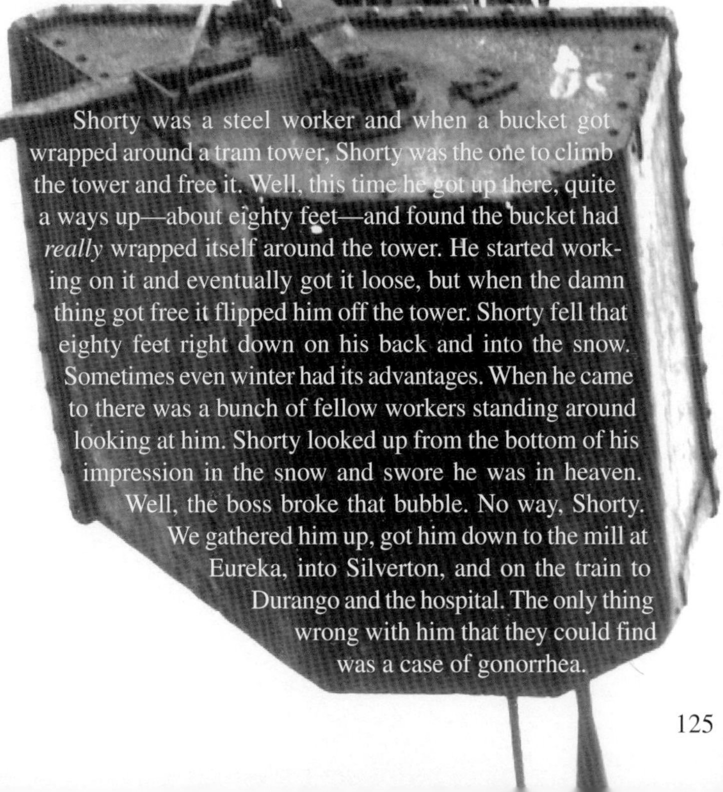

Shorty was a steel worker and when a bucket got wrapped around a tram tower, Shorty was the one to climb the tower and free it. Well, this time he got up there, quite a ways up—about eighty feet—and found the bucket had *really* wrapped itself around the tower. He started working on it and eventually got it loose, but when the damn thing got free it flipped him off the tower. Shorty fell that eighty feet right down on his back and into the snow. Sometimes even winter had its advantages. When he came to there was a bunch of fellow workers standing around looking at him. Shorty looked up from the bottom of his impression in the snow and swore he was in heaven. Well, the boss broke that bubble. No way, Shorty. We gathered him up, got him down to the mill at Eureka, into Silverton, and on the train to Durango and the hospital. The only thing wrong with him that they could find was a case of gonorrhea.

If you've got a mill that's almost four miles from the mine, then it's pretty clear you're going to need a tram. Above is the Sunnyside tramway in 1929. This span of cable is 2,000 feet long and 350 feet high. The ride to the mine took about 45 minutes. Tom Savich Collection

A wagon and team hauling tram cable to the mine by way of the Lake Emma road. This may have been George or Ernest Shaw leading the teams. The Shaw brothers were well known packers who lived in Eureka. They were famous men and were well educated. They went with Peary to the North Pole and had written books. They had won many contests in loading and packing mules and burros. Ruth Gregory is their descendant. Ruth Gregory Collection

Mickey Logan was an electrician at the mine in 1926. "The Sunnyside tram? Whoa, it was kind of shaky. If those buckets had another bottom I would've crawled right under. There was some wind in that country."

But Mickey remembers there were even less fun ways to get down. "I was at the mine when the boss came up to me. He says, 'Mickey, we've got a problem with the telephone. I want you to ski the line and find out what's wrong.' 'Okay,' I says, 'but it might be a little tricky 'cause I've never skied before.' I got out there and started down. Hell, pretty soon I couldn't even find the telephone poles. The snow was that deep. When I got to Eureka, I handed them boards over to the boys at the mill and told them, 'Boys, I don't care if you burn these darn things, my skiing days are done.'"

When Johnny Mac first saw the tram, it was almost too good to be true. He and Dude Gallagher would climb the first tower out, jump in a bucket, ride to Midway angle station and then jump out. One day they made a mistake. They jumped in a bucket occupied by Robert Salfisberg who proceeded to thoroughly thrash the two youngsters. The boys never did ride the tram again. That was in 1926. ❋

Looking down at the Sunnyside Mine and Lake Emma from Hanson Peak. Whiskey Pass is to the right (out of sight) and can take you over to Gladstone. The road down to Eureka is out of sight to the left.

John MacNamara Collection

The basic object of the tram was to bring men and supplies up the hill to the mine and to take ore from the mine down to the mill. Certainly the Sunnyside was no ordinary mine. It was Colorado's largest gold mine. (For a detailed history, read Allan Bird's *Silverton Gold.)* Ruben McNutt established the Eureka Mining District in 1873. He thought that by establishing this District, claims could now be legally staked on this once Indian-held territory. Waiting patiently until the next day, Ruben (who named Eureka) and his buddy George Howard (for whom Howardsville was named) set out to do just that. But the claim partnership failed to go smoothly, and shortly afterwards they parted company somewhat less than amicably and the confusing selling, swapping, splitting, trading dealings began. By 1877, Ruben's 'Sunny Side' had become one word, and continued to be used by the succession of owners including Engleman, Thompson and Judge John Terry.

However, no mine is an island. What happened around a mine, in the town and in the country, affected the mine just as much as the success of its interior workings. Trouble was always close by and in 1918 the flu came along costing Eureka fourteen percent of its population. The next year saw avalanches that took lives and knocked out the tram. Less than a week after these, on April 24, 1919, a fire broke out in the barbershop at the mine. The tram station, blacksmith shop and machine shop were the only buildings spared. Damage was estimated at $400,000 dollars. In 1919!

In 1920, due to plunging metal prices after the war, Sunnyside closed briefly, laying off 500 workers. But in 1921 the mine reopened because of finds allowed by the construction of the Washington Incline Shaft. The twenties proved to be good years, but by 1930 the Great Depression closed the mine and this time the town of Eureka as well. 1937 saw a brief reopening with as many as 380 men going back to work but by July 1, 1938, closure was pretty much complete and this time seemed final.

Quiet in the Sunnyside would exist until 1959, when, just after the shrine was completed in Silverton, Standard Uranium from Utah appeared. With the driving of the American Tunnel, lower extensions of the Sunnyside veins became accessible, and mining, this time from the old ghost town of Gladstone, once more became the mainstay of Silverton and San Juan County. But by the year 1994, the few miners left working at Standard Metals were no longer mining but were doing reclamation work instead. The tunnels that used to be so busy and productive, were being sealed with huge plugs of concrete. Once again, seemingly forever, Colorado's largest gold mine was on the verge of closing. ❋

Lake Emma and the Sunnyside mine buildings are shown in these two views taken from opposite ends of the lake. Try not to let the different photo angles confuse you. Zeke Zanoni Collection

LAKE EMMA AND THE SUNNYSIDE MINE

Note the tower in the center of the picture. It was put up to look for mine scabs coming over from Telluride in 1901 and 1902 when the Smuggler Mine introduced contract mining. The union objected and took to shooting the replacements. That tower was gone by 1937. Down below the tower lies the town of Eureka. The tram house is close to the tower, just out of sight. Below the tram house was a building that housed the steam boiler, fired by coal, that came up on the tram. That furnace ran continuously, three shifts. Between the two buildings was an incinerator that burned all the garbage and trash. Now, in sight, back up by the lake, the building right at the water's edge, is the pump house that pumped water to the mess hall and boarding house right behind it. Get out the magnifying glass—to the right of the mess hall are mounds of dirt. Inside those mounds were tunnels where all the perishables were kept. On the very right-hand side of the picture was another boarding house, pretty much just for miners as opposed to other workers. Between the two boarding houses sat the infirmary. There was even a doctor available. The long, low house to the left of the boarding house was the superintendent's place of business. Behind that on the right was the compressor house which housed three compressors, one of which was a backup. There were three continuous shifts here, yet inside this building it was as neat as a pin. The building to the left, back behind the superintendent's place, was where the blacksmith shop and machine shop were located. Now, look real close. In the back of the picture running almost the length of the photo you can see long snowsheds made of two by fours and tin. They ran from the mine portal way to the left over to the miners' bunkhouse on the right with various passages along the way. Quite a place! Ruth Gregory Collection

The stacks on the left are part of the steam plant. The coal came up on the tram to feed the boilers which steam-heated the buildings. They even had radiators in the tram house. The tram terminal is on the right. Note the power poles. The power lines run on top, the phone lines below. Some power came from Midway, generated by a water wheel. The rest came from the Western Colorado Power Company, supplier to Eureka. These pictures are from the collection of Gerald Swanson and probably taken around 1930.

The compressor room. Neat as a pin. There were three of them. The largest one was for backup while the two smaller ones were used every day. At least one man was kept working in here, cleaning and eye-balling the machines every day. It was a nice, warm place to be.

"When the wind blew, you'd put a blanket over the window in your room to help keep the snow out. God, it could sure get windy. But when the moon came out you'd swear you were in heaven."

The building in the foreground to our right is the same building as in the left hand picture.

This is the Sunnyside blacksmith shop around 1917. Walter Noll, a plumber for the Sunnyside stands on the right, watching Art Sullivan in the middle and Joe Sullivan with the double jack, making drill steel. They would get the tip of the steel red hot and Art would hold the forming hammer to the steel. Joe would swing the double jack, hit the hammer and make the bit. This hammer is in the San Juan County museum, donated there by Art, who was born and raised in Silverton.

Silverton was no isolated little town in the mountains. It was known world wide. Art was hired by the Russians to build an ore reduction and concentrating mill. He moved there with his family and spent two years supervising the building of this mill. Later life saw retirement in San Rafael, California on a farm raising fruit. John McNamara Collection

THE BLACKSMITH

So vital was he to the early day mining operations that the mine would literally come to a standstill without him. First and foremost he was the person who would keep the drill steel sharp for the miners who used them on a daily basis. But, beyond that he was the metal fabricator, making virtually everything and anything out of metal which was in constant need in every mining operation.

The larger mines had a number of people working in the shop. There was always the Master Blacksmith with his many years of experience and trade secrets which he seldom shared with anyone but his closest associates. He was the one who knew the critical art of tempering: drawing the one-point carbon drill steel into perfect condition. After the sharpening of the drill the tempering process would begin. With the re-heating of the bit edge to a cherry red, the steel would be dipped into a water bath (sometimes a mixture of fish or linseed oils was used), moving slowly in, then out, leaving enough heat in the body to start "the run." Using keen eyesight and experience, the master would watch the color start to climb the bit: blue, violet, red, straw and so on—the process taking only seconds. At just the right moment the steel would be dunked into the slack tub to stop the running of color; sometimes finishing cold, with hammer hardening on the anvil, when deemed benefical. If too hard it became brittle and would break while in the drilling process, too soft and it would dull rapidly and lose gauge too quickly. Many a miner cussed the "Greenhorn Smith" who thought he knew how to temper a fine piece of drill steel. Another worker in the shop was the tool sharpener who had advanced enough to forge the steel to the right temperature in order to pound out the cutting edges and establish gauge on the anvil. And of course there were the apprentices who where there to help, keep coke in the forges, swing the heavy hammers and learn the trade if caught up with the desire.

Even in the smallest mine someone had to sharpen and temper the drill steel on a daily basis. Sometimes in a moderately sized operation the boarding house cook was also the blacksmith. And, if someone on the property couldn't do it, then it had to be hauled into the nearest blacksmith shop to be done. But, done it would be, or you could take up farming for a trade.

Until the use of air-driven drills, the drilling was done by hand and called single or double jacking. The drill steel was to 7/8 inch in diameter with a chisel bit forged on one end. Normally, the holes to be drilled were twenty-four or thirty inches deep requiring four different lengths of hand steel from a "starter" about twelve inches long and ending with a "finisher" which was 2° to three feet long. Unless the ground to be drilled was relativity soft, the four pieces of steel were only good for drilling one hole, thus requiring four times the amount of holes to be drilled in steel needed for that days work. Twenty holes could demand eighty drill steel.

When using the early, large and heavy air-driven drills, which allowed from 4° to 5° foot holes to be drilled, the cross bit or star bit was forged on the end of the drill steel, requiring a different technique of drill sharpening. With the deeper hole came the requirement for longer and heavier drill steel and from four to six changes for a single hole depending on the type of drill used. Again one change of steel was only good for one hole thus requiring large amounts of drill steel to be hauled to the location of the drill for that days work. After the drilling was completed all the steel had to be hauled back to the blacksmith shop to be redone. It was not unusual for the larger mines to have any number of air drills in use at the same time.

You can see from the simple description above the amount of work that was required by the people in the shop. Only the larger mining companies could afford the newfangled air drills and they were cussed for years by the miner until they were perfected and supplied with water to keep the dust down. As for the Blacksmith Shop, it was open for business twenty four hours a day to keep up on the ever increasing demand of the mine. ❋ —*Zeke Zanoni*

At the junction two trammers (locomotive drivers) meet two miners pulling a flat car full of drill steel (just enough for one shift). Perhaps they were hoping for a tug to their destination. Zeke Zanoni Collection

ACCIDENTS ARE FOR KEEPS

Just like looking into the portal of the mine, the outside of the mill appeared to be a pretty peaceful place. But inside were many men doing many different jobs, facing many dangers. This story is reprinted from Snowflakes and Quartz, *by Louis Wyman—a day in the life of one man at the Sunnyside Mill at Eureka.*

The ambulance wailed like a Banshee with a toothache as it came up behind me, and I had to pull off to the side to let it go racing by. It was easy to imagine the scene inside the gleaming vehicle—the accident victim secure on a stretcher wrapped in blankets to stave off the chill of shock, well-trained medics standing by with oxygen, plasma and a hypo if needed. These men and women with their equipment are angels of mercy in this accident prone, speed-crazed world. I knew a time when I would have given anything for a little of their attention and skill.

Back in the summer of 1926 I had a job as mill-hand at one of the large lead-zinc mines of Southwestern Colorado. A good job for a youngster just a year or two out of high school. One morning, shortly after going on shift, I was badly mauled in an accident at the company's mill. The memory of the long ordeal I endured has not faded through the years. A steel brace I wear strapped to my back never lets me forget.

It happened a little before noon while I worked the table floor (the lowest floor in the mill), taking samples and oiling the machinery—just routine duty. One of my chores was the care of a long, overhead line-shaft, driven by a large electric motor. The shaft in turn drove twelve "Wilfley Concentrating Tables," by means of wide, rubber transmission belts. The installation was a constant source of trouble and demanded so much work and attention that I hated the thing. But it had to be maintained.

Working along from table to table, I noticed a stench of burning oil, always a sure warning of trouble up on the line-shaft. Over near the motor drive end a journal was smoking hot. In order to get to it with my oil can and tools, I had to scramble up an old two-by-four ladder. Mill superintendents and shift bosses took a dim view of shutdowns for repairs because of burned-out equipment. So I tried to adjust and lubricate the journal and get it to cool off while the shaft was still running.

Perhaps it was carelessness on my part, or just plain disregard for the danger. But somehow in turning to climb down I snagged my overalls right across the small of my back, and they started to wind around the shaft. Lashed to the shaft by my own clothes I would be beaten to a pulp in seconds; it turned at 250 rpm. For me, the chips were down.

In the split-second before it dragged me from the ladder, I grabbed for something to try and pull myself free and managed

From anywhere in Eureka you could see the Sunnyside Mill. Judge John Terry had the knowledge of how to build a mill. The first mill was built below the southern edge of Lake Emma, four miles above Eureka, in 1888. Wrong spot. The next mill was built by Terry at 'Midway.' The third mill and the first one in Eureka is at the left in this photo with the tram house at the top. And the last one was built in 1918, a large one, a 500 tons per day selective flotation mill, the first one of its kind in the United States. Only the concrete foundations stretching up the hillside remain today. *Ernest Hoffman Photo, Zeke Zanoni Collection*

to get hold of a water pipe with my left hand. Crazy with fear and pain, I fought that spinning bar of steel like a maniac, pitting my strength against the power of the motor in a grim tug of war. If I won I lived. If I lost I died. But manpower is no match for horsepower, and the strain almost tore my arm from the shoulder before it broke my grip on the pipe. Totally helpless I started to spin with the shaft's rotation. And I knew I'd never escape the whirling, smashing death just seconds away.

There happened to be enough space between the line-shaft and the cement wall so my head didn't strike as I made the first turns. My clothes began to rip away, and the thing peeled me out of them clean as a skinned rabbit. It hurled me against the cement wall with catapult force, and I fell to the floor twelve feet below. Even my socks were ripped off, although the white rubber mill shoes were still on my feet. I've never been able to understand how this could happen, but it did. Every stitch of clothing I'd worn formed a tight wad wrapped around the shaft.

Stunned and only partly conscious, I didn't realize I lay on the floor in the slime and muck, still alive. I couldn't move, get my breath or feel a thing. I didn't seem to have a body any more, just a head. And the thought came to me that possibly all but my head was a mangled mess wound up in the machinery. I wanted desperately to get up, just to prove I could stand on my feet. If I could do that, then my head and body were still attached. Finally I managed it and made a try for the stairs to the mill office on the floor above. But I staggered against a pulsating table deck. It struck me in the side and I went out cold.

A dim and bleary world greeted me when consciousness returned. The paralysis in my belly and chest had eased some. To be alive and lying there in the table spillage with a little air in my lungs was a wonderful relief. I crawled over to the steps and sat, trying to take stock of myself.

Then the pain came—so intense it made me sick. From my shoulders and back it seemed to flow in sheets to every part of my body. My left arm hung useless from the shoulder, and it had swollen so badly it scared me. My legs were not much better off either, but at least I could move them. Blood dripped from my nose, mouth and a dozen deep lacerations. I kept trying to get things in sharper focus not realizing both eyes were black and almost closed. A lot of skin had been burned away by my clothes when the shaft stripped me. Slowly I began to understand, I'd been seriously injured.

Calling for help was wasted effort—but I tried. No one could hear my voice above the crash and roar of the mill. An hour might pass before the shift boss came by on his rounds. I had only one out: climb the stairs to the office on the floor above for help.

I rested a few minutes and then faced the long flight of steps and started up. I panicked with the thought of falling back down again but kept telling myself, "Just a few more steps and you'll have it made." Those few steps took a lot of doing.

Finally when I staggered through the office door, the mill superintendent got the shock of his life. He all but fell over backwards out of his chair. "My God kid! What happened to you?" he yelled. Then he started firing questions without waiting for an answer. After he'd calmed down a little, I tried to tell him, but I don't think I made much sense. He phoned for a relief operator to come out and take my place. The operation of the mill was always uppermost in his mind.

Louis Wyman, on the left, was born in Silverton in 1901. The next year his father, Louis Senior, built the Wyman Hotel in Silverton. Young Louis graduated from Silverton High School in 1921. He was working at the Sunnyside when the Casey Jones was built (see page 136). His mother Ellen and his brother Robert are beside him at their house at the end of Reese Street, circa 1940. Grace Besaw Collection

Chris, my shift boss, came in about then, and I had to tell him what I thought had happened, all the while standing naked in the middle of the floor. He and the mill super just stood and gaped at me in astonishment. They couldn't think of a thing to do, and I was getting pretty light-headed. Chris came out of it first and rustled up a pair of old bib overalls and a jumper. "Here put'm on kid," he said. "You look like the wrong end of a butcher shop." I couldn't get into them so he helped me and then took some wet wiping waste and did what he could to clean up my face.

They thought it best I take the stage to town and have the doc at the company hospital look me over. Chris walked with me to the main office to make sure I got there. "You better wait here, Kelly will be along with the stage soon," he said. "And have the doctor take a good look at those cuts, as well as your arm." Chris figured he'd done all he could for me, and hurried back to the mill.

I didn't want to go into the office and sit down. I didn't want anyone to see me, so I waited outside and held onto the handrail to keep from falling. By the time Kelly pulled up in the old touring car he used for a stage, things were getting pretty vague. But that helped a lot as I eased myself onto the front seat. And I prayed I'd never have to move again.

Kelly was everyone's friend, as well as mine. "What happened to you?" he asked. "You fall off of the mountain or something?" I couldn't tell him much about the accident, and it miffed him a little. He took a long hard look at me but didn't say anything more and went about his business, delivering the stage freight to the mill. Then we drove over to the post office for the outgoing mail. A time consuming task. Kelly never pulled away from a stop without a complete run-down on the local gossip.

After the post office he headed for the company boarding house. "Come on in and have a cup of coffee, you look like you need it, and it's on the house," he said. "I'll have to see if the cook wants anything from town." I stayed in the car trembling uncontrollably from chills and shock. I wanted a cup of coffee like nothing else in the world right then. But the price I'd have to pay in pain to get it was too high.

When Kelly finally had all his details attended to, we went rattling down the canyon road to town. A long nine miles away. He did his best to miss the rocks and chuckholes, but I know he hit every one of them. Nauseated and weak, I chewed my lips, squinted my eyes and hung on. Shortly after we'd made a stop at a small hamlet to pick up the mail, Kelly redeemed himself. He gave me a drink of good bootleg whiskey from his bottle. It was a julep from heaven.

We made our last stop at a crossroad post office, and when Kelly came out with the mail sacks he said we had to make room for a lady passenger. He guessed I'd better get in the back seat. Somehow with his help I made the switch and climbed into the tonneau. It took a little time for the lights to stop flashing in my eyes and for me to quit swearing from the pain of moving about. Kelly hadn't mentioned the lady lived up a side road a mile or so.

Our passenger came out to the stage wearing her finery and a big smile. A trip to town on a bright summer day was no small event. Kelly, the gallant, grinned his welcome and held the door for her. She glanced up and saw me. Stifling a scream, she backed away from the car. I must have been a sight perched up on the freight and mail sacks, dressed in old dirty overalls, both eyes black and blood oozing from my nose and mouth.

Trying to wipe it away only made it worse. If Kelly hadn't assured her I was harmless I think she would have run back to the house. Poor woman, I spoiled her trip. She sat on the very front edge of the seat, as close to the dash as she could, all the way to town.

Kelly brought the stage in late as usual. He'd run out of road, but not errands. First he saw the lady to her destination. Before he could open the door for her, she'd taken another look at me and fled. Naturally the post office was the next stop. It had priority and was the terminal for the stage line. "No need for you to walk over to the doctor," he said. "I have to drive by that way, and you may as well ride." A very friendly and considerate gesture on Kelly's part. He let me out in front of the hospital, and drove off.

From the sidewalk, I looked up at the second flight of stairs I'd faced that day. These were wide and not so high. But the steps seemed to be moving up and down in waves, and there was no handrail to hold onto. I wondered how in the world I'd climb up. I thought perhaps, if I spoke to them softly as a person would to a frightened or excited animal, they'd hold still. It worked, and I climbed up to the door and made my way inside.

Later one of the boys who was "Goldbricking" in the hospital, said a nurse found me standing in a hallway corner, chattering and shivering like a scared monkey. The pain in my body had been building in intensity all the time. I was far from rational, and I didn't want anyone near, for fear they might touch me.

I received the first pain killing shot three and a half hours after the accident. And when they'd finished with the necessary stitching and patching, I resembled a mummy more than a man. I couldn't walk or move without help. They'd strapped me to a plank, and I learned to know that piece of lagging very well indeed.

My score for the day was a broken back, a left arm all but yanked off my shoulder, some badly cracked ribs, and several joints that still give me some trouble. Sewing up the cuts, the bruises, and the skin I lost don't count. In time they built a steel frame to keep my back in line with my hips and shoulders, and I get along fine.

And friend, you there in the racing ambulance, too bad you were hurt. But since you were, you're in luck in that vehicle. There was a time equipment like that and trained personnel to man it didn't exist. For the unfortunate the going was rough. One thing hasn't changed much though. We still learn the hard way, that "accidents are for keeps." ❋

There had been an old 1915 Cadillac sitting idle at Eureka for at least two years that had belonged to the Sunnyside. Then the boys at the mill got hold of it and turned it into the 'Casey Jones.' Henry Gray became the operator and drove it between Silverton and Durango until the mine closed. Eventually the railroad said 'no more using our tracks' and the car returned to Silverton where it rests today next to the museum. This picture was taken in the forties when the machine was parked at the end of town by the present day lumber yard. *Gerald Swanson Collection*

THE CASEY JONES

"The Sunnyside manager, in one of his more civic-minded moments, ordered machine shop foreman Casey Jones to build a motor car that could operate on the Silverton Northern railroad tracks. The idea was to provide a fast means of transportation to the hospital in Silverton, should men be injured in an accident. Of course, there were some fringe benefits. The car would be a handy means of travel for management personnel on company business or pleasure, or for Eureka residents attending lodge meetings in Silverton. So Casey and his boys went to work and built the famous railroad car, 'the Casey Jones,' still on exhibit in Silverton at the San Juan County Historical Society museum.

"A host of men worked on constructing the rail car under Casey Jones' direction. Among them were Bill Jones, Clyde's brother, who was working out his apprenticeship; machinists Horace Sease, Earl 'Speedie' Jennings and George Gallagher; blacksmith Ben Caine; Millwright Hans Tansted; and carpenter Lars Larson… While the car did occasionally function as an emergency vehicle, more often it transported party-goers or lodge members to Silverton and back of an evening." *From Dusty, Citizen of Eureka by Louis Wyman.*

And the party-goers just weren't the people going from Eureka into Silverton. Gerald Swanson's mother Mary, Julia Maffey, and Lydia Carter rode out to Eureka around 1920 to go to a Christmas dance. Old Casey was quite the social vehicle.

But just as parties and time of revelry were only a small part of a life not destined to last, so, too, the mines were destined to lives of heady production followed by closure. By July 1, 1938, the Sunnyside mine closure was pretty much complete and this time seemed final.

When the mine closed the town closed. In Eureka the miners left. The houses were abandoned. In 1942 the railroad tracks were yanked out. But in Silverton there remained a town full of enterprising people.

It was in the forties that houses from Eureka started showing up in town. Sometimes the whole house, sometime bits and pieces. One enterprising young gentleman bought a house for ninety dollars and was going to part it out as the stepping stone to his first fortune. Today, his garage is roofed and his porch is sided with Eureka pieces, but that first fortune proved most elusive.

Not all of the houses just went to Silverton. And not all left Eureka in pieces. You know those buildings you see

The Longstrom home as it appears in the 90's. One door shuts and another opens. *John Marshall Photo*

Carl's mother Mary Longstrom, lived with her mother in Silverton around 1920. *The Longstrom Collection*

on the other side of Red Mountain by the Idarado? They came from Eureka.

Carl and Joan Longstrom tell of their home's voyage:

"Our house came from Eureka on telephone poles, pulled by a truck. It had log rollers and two cables running across underneath, holding everything together. If you were to look under the house today you'd see that the rollers are still there and so are the cables. When they dropped the house off though, the front was in the back and the back was in the front. So they looked at that and decided to chop off the front, which was really the back, and put on this little brick addition. We love the place and hope to move in full-time soon. Oh yes, we think there may be another half to our house somewhere. Maybe our neighbors…" ❋

Carl and Joan Longstrom generously share pieces of their past. Carl spent his life in the film industry. The first movie he worked on was 'Ticket to Tomahawk' in Silverton in 1949 and the last one was 'City Slickers' in Durango, some forty years and a Who's Who list of stars later. *John Marshall Photo*

Ruth Gregory Collection

Never mind the steep railroad ride. Forget the rockfall, the ice, the avalanches. Don't let the rocky road bother you. It's the nineties, the railroad is gone, but the road's in pretty fair shape. At Animas Forks the few buildings left stare at you through paneless eyes, the columbine bloom and the air is crisp and fresh and it is so quiet.

Good-bye Eureka Hello! Animas Forks

The town of Animas Forks as it appeared in 1954. The Gold Prince Mill would have been in the lower left of the photo. Note the Bagley Mill in the background on the right side up on the hill. N.R. Bagley, in connection with the Rockefellers, was responsible for backing the Frisco Mines and Tunnel Company.

Ruth Gregory Photo

This wooden building, reroofed and still standing in Animas Forks today, is known as the Walsh House. It was built in 1879 by miner and mailcarrier, William Duncan. *Ruth Gregory Photo, 1954*

There is a certain danger in leaving Eureka and heading the four miles up to Animas Forks. The danger is going into such a once busy and active place and only touching on so little. Every valley around the town of Silverton abounds in its own history and stories.

In 1885, 450 souls lived in Animas Forks.

The Silverton Northern train reached here in 1904.

The *Animas Forks Pioneer,* published in the year 1882, was the 'highest' newspaper ever printed and published in the United States of America, at approximately 11,200 feet. From here one could travel across Engineer Pass, Cinnamon Pass, or up Placer Gulch and California Gulch. Some of the mines were: The Vermillion, The Sound Democrat, The Mountain Queen, The Gold Prince and The Frisco Tunnel. ❋

The huge structure of the Gold Prince Mill pictured in this photograph is only echoed today as a series of foundations stepping down to the river. It cost $500,000 dollars to build in 1904. It was made of structured steel and, after running only six years, was dismantled and recycled in the new mill at Eureka. The train reached Animas Forks also in 1904, no small feat—climbing 1,205 feet in 3° miles. The grade averaged 7 to 7° percent and proved to be the very maximum for the steam railroad. *Ruth Gregory Collection*

There always seems to have been an unaccountable aura of romance and mystery associated with mining and the lure of GOLD. That aura apparently was able to turn the heads of many shrewd businessmen located far away. Two years after the Gold Prince Mill shut her doors men were hard at work, supported by far away money, endeavoring to build another mill only a mile away. Sometimes such efforts paid off handsomely—more often they did not.

THE BAGLEY TUNNEL

Gregg Harlow of Broken Arrow, Oklahoma pieced together this chain of events from articles published in The Silverton Standard and the Miner, *Colorado's oldest continuously running newspaper:*

December, 1903–Property belonging at one time to the Mountain Mining and Milling Company, consisting of the Red Cloud and Vermillion groups plus the territory in the immediate vicinity comprising some 140 claims in all, are to be carried out through one tunnel near Animas Forks where connection is to be made with the Silverton Northern Railroad. N.R. Bagley, in connection with the Rockefellers, is backing the enterprise. The properties were worked in 1870's and produced 800 tons of ore. At that time the nearest milling point was Pueblo, Colorado, with transportation charges of over $40 dollars per ton.

August, 1904–Frisco Mines and Tunnel Company president and general superintendent of mine and operations, N.R. Bagley, was in Silverton this week and informed the paper that ten miners are now employed and working on the properties "that embrace one of the largest mineralized territories under the management of one company in the San Juans." Just at this time three of four cars of ore from the Red Cloud and adjacent mines are being readied to ship to Durango for smelting as soon as the railroad extension from Eureka to Animas Forks is completed.

June, 1905–The company is incorporated for $6 million dollars. It is proposed to build a railroad tunnel through Houghton Mountain in the San Juan Valley (California Gulch) to Poughkeepsie Gulch. The cost is estimated at $500,000 dollars for a total length of three miles. The directors of company are William E. Morgan, Harrison H. Merrick, James M. Sutherland, William E. Walker and N.R. Bagley.

October, 1905–A contract was let with Rich Whinnerah (Silverton) for $100,000 dollars for 5,000 feet of tunnel. It will also be used (tunnel and railroad) for the transit of passengers and freight for other mining companies in the district. Just now men are installing compressor and drills of the latest type (Ingersoll-Sargeant) and making preparations to work sixteen men three shifts during the winter. Mr. Whinnerah is confident he will, with such splendid machinery, be able to make over two hundred feet of tunnel per month. The Gold Prince Mill which is in close proximity has a capacity of 500 tons. The Frisco people will be beginning operations with the Gold Prince Mill.

October, 1906–The tunnel is now 1,000 feet under Houghton Mountain and at the breast has a vertical depth of 900 feet.

April, 1907–The Bagley Tunnel is in nearly 4,000 feet and is making ten feet a day. The electric drills have been discarded and three big old Sullivan sluggers are doing the work.

Richard and Violet Perino. Mining historian Rich Perino has preserved a wealth of information about early mining in the Silverton area. Rich shared with me this strange history of the Frisco Mines and Tunnel Company which was compiled by Gregg Harlow. Richard, himself, worked for Standard Metals beginning in 1961 and finally called it quits in the layoffs of March 15, 1985. He first worked at the mill, proceeded to the mine shop, and from there went underground doing just about everything at least twice until his retirement. That is, his body retired but not his mind. It keeps right on mining and throughout this book we are grateful for his valuable contributions. *John Marshall Photo, 1994*

November, 1907–The tunnel, when completed, will be 13,000 feet long and is now nearing the 4,400 feet mark. It is being dug straight as an arrow.

November, 1907–The Bagley Tunnel is being dug at a rate of eight feet a day. The contractor has recently installed Water-Joyner drills.

June, 1908–During the month of May, 203 feet were added to the length of the tunnel which is now 5,500 feet.

July, 1909–A party of surveyors have been at work the past few days laying out a mill site near the portal of the great bore.

April, 1911–A short distance above Animas Forks is the great dump and portal of the Bagley Tunnel. It is now 7,500 feet; dimensions are 7 by 7° feet in the clear…other tunnels of minor development are being driven.

September, 1911–George Eisenbois, representing stockholders of the Bagley Tunnel, arrived Tuesday evening and spent considerable time with manager Gagner examining the workings of the great bore. He had the pleasure of looking upon one of the finest pieces of tunnel work in the whole district. The present amount of ore opened up in the levels driven from the main tunnel fully justifies the construction of a 150 ton mill.

February, 1912–The erection of the proposed mill at the Bagley Tunnel will bring to light and action one of the greatest mines in the state. It is one of the best pieces of work ever performed in the whole district, with not a curve or crook for the entire distance of the tunnel, and when the electric lights are on in full from one end to the other it is a most beautiful sight.

April, 1912–Already the work of framing the timbers for the new mill to be erected at the Bagley Tunnel has commenced and is being done at the San Juan Lumber Company's yard under the supervision of Paul Hansen and M.L. Porter. With the erection of this 150 ton mill… manager Gagner will push work ahead to complete this mill as soon as possible, and before the close of the summer months the new mill will be going at full speed.

June, 1912–Work on the framing of the great mill for the Bagley Tunnel is progressing rapidly and it will be ready for shipment by the time the road is open to the Forks.

August, 1912–The main structure of the great building is growing rapidly and the roof is now being put in place. Most of the machinery is now on the ground. The general character of these veins are somewhat varied… The tunnel itself is one of the most perfect mining engineering feats and works ever performed in the San Juan district… It penetrates the center of Houghton Mountain on an even grade and is as straight as an arrow. The whole working system is lighted by an electric light system, with powerful incandescent lights, at frequent and regular intervals, and when the mine is in full operation it presents a scene as enchanting as a fairyland picture. *However, as beautiful as it was, it soon became another project which comsumed a vast amount of money with little or no tangible results.* ❋

The Bagley Mill of the Frisco Mines and Tunnel Company was initially framed at the San Juan Lumber Company in Silverton and completed in place in the summer of 1912 just outside of Animas Forks. Back on page 138 you can see the mill still standing in 1954. Parts of the building remain today. Gregg Harlow Collection, circa 1915

About a mile above the Bagley Mill lay the Mountain Queen, one of the early mines of the area and one of the least known of the good producers of the San Juans. The Mountain Queen ran, albeit sometimes raggedly, for almost eighty years. The following text is a brief and little known history of the mine, a history so typical of all the mines in these mountains.

THE MOUNTAIN QUEEN

Found at the extreme end of California Gulch on the north slope of Hurricane Peak west of Animas Forks, the Mountain Queen was located in 1877 and brought to patent by F. Wheat, James Fairchild, Joseph Merrill and Charles Tall in March 1884. At the time of the mine's discovery silver was the name of the game and the owners knew they had a winner with the gray copper assaying over a hundred ounces of silver per ton on the outcrop. But like other mines of the time, in order to be profitable its ore had to be hand sorted into the best product possible since it had to be transported by jack (burro) or mule train over Cinnamon Pass to Lake City and beyond. This left large amounts of base metals on the dump which later became profitable when civilization finally caught up with the rush to the San Juans. But until then the going was rough, not only in the shipping of ore but in getting the huge amounts of food and supplies which were needed to sustain the operation.

The first recorded shipment of ore from the Mountain Queen was in the fall of 1877 being somewhere over 300 tons. This shipment went to the Crooke Smelter in Lake City. The following year another shipment of 500 tons to Lake City was recorded. Although these shipments seem small by today's standards, they were good for a mine breaking rock with hammer and handsteel and working under the conditions which existed at the time. One form of measurement used for this type of mining was one ton of ore per day per man employed at the mine. At times the average was only half that. Of course the tonnage per man increased as air drills were employed.

Typical of most mines at high altitude, the Mountain Queen sank a discovery shaft on the outcrop of the vein. With the shaft down to almost 200 feet the high assays of surface silver diminished somewhat, but the vein widened, producing large amounts of pure galena (lead) with other base metals along with silver and some gold. The grade of ore was excellent, but keeping the pack trains on the go was always a problem. Fighting the chill of high altitude in the summer was tough enough, but keeping the trails open after the first snowfall in the fall was never easy. Man and animal suffered alike.

By 1880 the entire San Juan Mining District was booming, hundreds of miners and prospectors were flowing in, building towns, mills and smelters. What more could one ask for? It was exciting for those with the producing mines. Then in October of 1880 Professor James Cherry, who had just completed building the Eclipse Smelter three-quarters of a mile below Animas Forks and in need of ore, bought the Mountain Queen for $125,000 dollars along with several other properties in the area. With the heavy winter of 1881 the Mountain Queen along with most other mines had to call it quits. Come spring "every man that was available was immediately employed to shovel out the roads and trails."

Everyone knew the railroad was coming. It was almost to Durango, and by 1882 it would reach Silverton. If there was any one thing that changed the face of mining in the San Juan, it was the coming of the railroad. Not only was the transport of ore and mill concentrates cheaper and faster but it allowed the shipping of lower grade ores that laced many of the mine dumps. Heavy machinery and supplies were readily shipped into Silverton, and even though everything still had to be packed up to the mines, it was a far cry from the long and expensive trip through the mountains and on to Pueblo. From the summer of 1882 on, the day of the pioneer would be gone for ever.

By the summer of 1881 the Eclipse Smelter was in need of ore and lots of it. "A contract was let out in the Mountain Queen of taking out, sorting and sacking 1,000 tons of ore at five dollars per ton. Also to sink the shaft another two hundred feet. First one hundred feet by hand winze, second one hundred feet by power. They will continue drifting on the Red Cross. The Eclipse is also delivering lumber to the Mountain Queen and Red Cross for boarding houses, ore sheds and blacksmith shops. In these various operations, Professor Cherry has employed one hundred men, twelve four mule teams and one hundred burros."

By March of 1884 the ownership of the mine was becoming somewhat confusing, with partners selling shares or parts of shares to new partners. An example of this was Norman Bingham selling one-twelfth interest in the Mountain Queen Lode to Arthur and Horace Brock for $3,666 dol-

A mule train has reached the Mountain Queen, around 1927. **Gerald Swanson Photo**

lars. Then immediately following this, Joseph Merrill, a major partner, sold one-quarter of 14/15 interest to the Brocks for a sum of $20,000 dollars. This kind of thing along with names being dropped and/or added to the record makes detailed research nearly impossible.

By 1885 the management of the Eclipse and Mountain Queen had passed from Cherry to Professor S.W. Thome and the Articles of Incorporation on the Mountain Queen were filed in the county courthouse. With the Eclipse in financial trouble, Thome was doing everything he could to save money and one big expense was wages. With this the sparks started flying at the Mountain Queen. On August 29, 1885 the *Miner* came out with this article, "There is likely to be trouble at the Mountain Queen mine before many days have passed, arising from the fact that Mr. Thome, the manager, has cut down the wages of the men to $3.00 and $2.50 per day. The men number about twenty, but being in needy circumstances, are content to work until they make a 'stake,' where they will go to the other mines where the regular pay is $3.50 per day. Other miners of the Animas Forks and Mineral Point areas resent this, arguing that if the mine is allowed to work at this cut rate, it will not be long before every mine in the county will adopt the scale." The article went on to say that "The Mountain Queen men want to work and management will not listen to the outside miners. There has been a meeting of about 150 local miners to devise some means whereby the men on the Mountain Queen may be dislodged. These outside miners are growing desperate, and may resort to force, unless the Mountain Queen crew quits. Most of the miners hope no violence will be needed, they feel that the talk of using giant powder to blow up the mine should be tabooed and the instigators fired from the meeting. The men will thus gain the sympathy of the community with things coming to a natural course of events." Unfortunately there was no follow up article to the event, but apparently the labor dispute was settled since the Mountain Queen continued to work, being one of the better producers of the county for years to come.

Once he'd had an interest in the Mountain Queen. Tom Walsh sits in his office at the Camp Bird Mine outside Ouray. No fax, no computer, yet he made millions here. Imagine! *Ouray County Historical Society*

In October of 1885 another article appeared in the *Miner*, "Papers have been filed this week transferring all property of the Eclipse Mining and Smelting Company to the Mountain Queen Company, this move being made to settle some litigation." A week later another article states "Professor S.W. Thome, general manager of the Mountain Queen Company, left Tuesday on the train for Philadelphia." This is the last we hear of Thome. But problems at the Mountain Queen must have continued between the men and management in late 1885, running into 1886, causing a shutdown of the mine. The *Democrat* came out with an article in October 2, 1886 stating: "The chances of the Mountain Queen resuming work very soon are growing brighter if the company can come to a satisfactory agreement among themselves."

Things apparently came together, because by the summer of 1887 Professor Ihlseng, returning from Chicago, was advertising for bids to sink the shaft another 100 feet and buying new machinery. By fall of that year they reported that the machinery was in place, they were making good progress in sinking and were making regular shipments of ore.

Following a few newspaper clippings on the progress of the mine, *The La Plata Miner*, in August 23. 1901, came out with this: "The Mountain Queen has made a rich strike. An ore body has been uncovered for a distance of over seventy feet which at times is six feet wide. It contains a streak of solid galena varying from twenty to thirty inches wide. The galena is being shipped right along. Last year considerable ore was shipped, the lowest grade being fifty-three percent lead. Assays on the new strike average sixty-five percent lead and thirty ounces silver, and a small amount of gold per ton. "Ironically, in July 1892, less than one year later and still a year before the 'Silver Crash of '93,' the Mountain Queen Lode was sold by the county for three years back taxes. For a total sum of $662.65, Edward P. Griswold, also of Chicago, walks away with full title to the mining lode. What happened to Merrill and the Brocks' investments only eight years prior of over $25,000 dollars? Here's a mine in production one year and going for taxes the next, giving the impression that everyone just quit and walked away. This was also a mine that sold twelve years earlier for $125,000 dollars and thirteen years in the future will sell again for $100,000 dollars as you will see. Yet, nothing more could be found on the previous owners!

Many of the mines with a reputation of being a good producer in later years eventually went the way of leasing, and the Mountain Queen was no exception. By the early 1890's Thomas F. Walsh (who in the near future will be famed as the owner and developer of the fabulous Camp Bird mine near Ouray), referred to by some as a "Dump Robber," held working leases on the Mountain Queen, Vermillion and Ben Butler mines among others near Animas Forks. He worked the Mountain Queen a number of years sending the ore directly to his smelter in Silverton. His living in Animas Forks as well as Silverton during this period may have given rise to the name 'Walsh House' in Animas Forks to the house built by Duncan. Since the 'Old Timers' of the region always referred to the house by this name, there is a good chance that Walsh did live there for a time, for the name has stuck up through recent times.

Walsh, at the time, was having financial problems in this period of his life, trying to make ends meet on the number of mines he was leasing here and in Rico. Taking from the book *Father Struck It Rich* written by his daughter Evalyn Walsh McLean, Walsh was unable to make the mine pay off, and to his friend he wrote, "The Mountain Queen pay streak does not hold its value going down. For 600 feet on the surface it averages 100 ounces (silver). Four feet down it falls to thirty ounces. On account of snow I have been unable to work the Vermillion yet." He later continued, "I have made no test on the Mountain Queen as a concentrating proposition as I have no means. I paid in the early spring an assessment of $250 on (Ben) Butler. We are getting some good ore from this mine but expect it to play out any day." The letter went on to say "Within myself I feel—even though no one else agrees with me—that I recovered victory from defeat at Silverton (the smelter), made what was a lost investment worth something."

He wasn't the only miner having problems, the 'Silver crash of '93' shook the very foundations of the Silver San Juan. It would take a few years of better milling practices for base metal recovery and the discovery and development of the gold bearing veins to make the economy rebound. And rebound it did, for Silverton and San Juan County would see their best years at the turn of the century and beyond through its first decade.

In August 1904, Rasmus Hanson, a highly respected mining man and one who will be famed in the future for the successful development of the Sunnyside Extension and Mastodon Mines, acquired a three year lease-purchase on the Mountain Queen from the estate of Edward P. Griswold. A year later he was given a one year extension with the agreement reading in part that he was to pay twelve percent in royalties minus shipping and that the royalties could go towards the $25,000 dollar purchase price if paid within the three year period. The last payment of $10,000 dollars to be made on or before May 15, 1908.

Supplies were hauled by a wagon with runners when the snow was on the ground, which it usually was. Harley Short was the freighter in 1928. *Louis Girodo Family Collection*

This deal certainly proved the business capabilities of Rasmus Hanson, for in that very year (1905) he turned around and sold the Mountain Queen to a group of Cripple Creek capitalists for $100,000 dollars, giving Rasmus a tidy profit for his work. Shortly following, the Guggenheims, who just a few years before had purchased the Silver Lake and Iowa Mining companies in Arrastra Gulch, also secured an interest.

At this time Fred Bodfish as a partner also become General Manager. And the company apparently meant business because they immediately installed a 'four-drill' air compressor and boiler. Up until this time all mining had been done with hammer and handsteel but quite successfully since the Mountain Queen had been credited for $250,000 dollars worth of 'smelter rock' (ore rich enough to eliminate the milling process) from 1877 to the turn of the century. This did not include the lower grade ores that had been left in the mine or piled on the surface dump.

Yet it was on March 17, 1906 that eighteen men were killed by snow slides running throughout the county. Twelve were killed at the Shenandoah-Dives bunkhouse. Another man, a watchman from the Sound Democrat, was killed over in Placer Gulch right near the Mountain Queen. He had spent the night at the Sunlight Boardinghouse, where he thought he would be safe, instead of at his own. Yet the mining went on.

By March of 1906, the new owners were running a tunnel along the principal vein to cut the old shaft. The Mountain Queen tunnel extended 1,200 feet, coming under the shaft (which by now was 430 feet deep) where they raised up to make the connection. By this time 15,000 tons of $15 dollar ore was piled on the dump "which will be marketable at a good profit under the modern process of treatment." In August of that year the *Silverton Standard* reported another rich strike "which resembles closely the ore taken from the Old Lout…It carries a heavy percentage of bismuth and runs over 200 ounces in silver and from one to four ounces in gold." Two months later the Standard reported "Manager Bodfish, fully convinced that he has opened up a wonderful mine, is now erecting new buildings and making all necessary arrangements for a vigorous winter campaign." In December of 1906 the company also installed a new 10-drill Rand compressor, increasing their rock breaking capabilities.

And a vigorous winter campaign it would prove to be as the next *Standard* account reported in March 1907. With the first part of the article giving a progress report on the mine including a newly encountered ore body, it finished with the following: "It is a record-breaking feat for the San Juan, to have the power plant taken out by a snow slide in November and a new plant installed and set to work in March, but this is what Manager Bodfish has been able to accomplish, and only shows the energy and vim with which the development of the Mountain Queen is being prosecuted. The new boarding house which was recently completed is pronounced one of the most comfortable and up-to-date in this section, being heated throughout by hot water and comfortable in every particular. It is a first class hotel, with all the modern comforts, at timberline."

In the summer of 1907 the management of the Mountain Queen was talking of erecting a mill one mile down the gulch, connecting it to the mine with an aerial tram, to handle the large amounts of lower grade ore which was being continuously produced along with the 'high grade.' This was never built, probably due to a favorable milling contract with another party, saving the huge expense of building their own.

The new Mountain Queen Mining and Milling Company which was incorporated in 1906 continued to operate, or at least maintain possession of the mine with varied success until 1912 where the entire group of claims reverted to San Juan County for non payment of taxes. Three years later the group of claims were taken up by the Moffat Estate Company in 1915, who conveyed them to the newly formed corporation (1916), Eureka Mining and Milling Company in 1917. This corporation has maintained control of the mine granting various leases through the years up to the last known lease in 1956.

By 1927 Art Walker would step in and obtain the lease with two other partners. Art and his wife Cecil worked the Mountain Queen leases until the late 1940's. Out of Silverton, Rich Perino and Bill Maguire Jr. both trucked ore from this mine to the Mayflower Mill, thus closing an almost eighty-year mining history on one of the least known good producers in the Silver San Juans.
—Zeke Zanoni

By 1989 the boiler was pretty much all that was left to denote where the Mountain Queen had once stood. Cynthia Francisco and Don Spencer stand beside it. ***Don Spencer Collection***

(top) Fury Dalla, with the thick head of hair, stands near Joe Todeschi's father, Albino (pronounced Albeano), who is swinging a hammer. Albino was the blacksmith. Art Walker leased the claim. Rock would have been loaded on the bare boards, sorted by hand, bagged and loaded out by pack train. Gerald Swanson Collection, 1927 or 1928

"If you look closely there, the steam plant is to the right, behind the building. Yes, that's my father there, the man on the right. I'm sure of it. Albino Todeschi was the blacksmith at the Mountain Queen in 1928. The crew had worked all winter. Harley Short, a well known San Juan packer, brought in the supplies like powder, groceries, timber and coal in the fall. The following summer, they packed out the ore. Silver—you bet. It was enough to load two rail cars. The first car was worth $18,000 dollars, the second, $24,000 dollars. I was twelve years old, but I remember me and my sister Cecilia (who married John Troglia) rode our horses up to Animas Forks for a picnic celebration. We left Silverton at 6:00 a.m. There must have been forty people up there. Food? We didn't go for the food. We went for the soda pop. I think I got away with a pretty good size bag full of the stuff. Cecilia rode the train home. That was a good mine and a darn good picnic." —Joe Todeschi, 1994.

These rocks were all 'hand cobbed.' An interesting mining term from Cornwall. In the 1700's one would use a cobbing hammer to remove as much waste rock as possible. By hand. The rock that left the mine would be sorted before going outside. Then on the surface the best ore would be sorted again, the fines going out bagged and maybe the bigger rock would be shipped by wagon. The purest ore would go directly to smelters, probably in Pueblo, the lower grades would go to a mill.

(bottom) In this photo a lot of ore remains to be bagged but there's a lot ready to go out. Were there enough men working to have their own cook? Apparently so. Those clean white clothes sure stand out. Louis Girodo Family Collection, 1928

The difficult section of rail from Eureka to Animas Forks never did pay for itself. Regular operation ceased fairly soon after its arrival although apparently trains did run occaisionally up to Animas Forks until the twenties. Otto Mears told the county they could have the right of way if they hauled away the track. By 1936 the tracks were gone and the road was what it had been before the rails. Let's wander back down the road to the train in Eureka which ran sporadically until at least 1939. Let the years drift by and head on into Silverton. No need to hurry. ❄

The sorting floor is bare.

(above) "I was pointing out the trail to Lake City to Mr. Blick when Pelé took this picture." The sorting floor is covered with ore.
(below right) This is the grizzly that would shake out the fines (the smallest ore).

The bags of ore build up by this unused-looking house in Animas Forks, awaiting shipment to Silverton and on to the smelter. *All photos from the Louis Girodo Family Collection.*

In fact, while we're taking our time let's look around the town of Silverton a bit. The following article helps us to understand some labor struggles of the miners. We can see in the following pages as well, the industrious and creative efforts that were undertaken to build lasting structures—physical, familial, commercial, industrial, entertainment and communications— which are the foundation of community in Silverton.

Silverton Miner's Union

A YEAR AFTER THE *SILVER PANIC OF 1893,* when the United States government demonetized silver, the Silverton Miner's Union Local #26 was formed by the Western Federation of Mine, Mill and Smelter Workers to counter lowering of wages and increased working hours by mine owners. The fortunes of Silverton had depended heavily upon the production of silver, particularly from the rich Red Mountain Mining District, and the steep fall of its price caused dramatically reduced profits for the mine owners, who looked for ways to cut costs.

By 1907, however, with the increased production of gold as well as silver, the mining industry in San Juan County had stabilized. The population of the county was close to 5,000 and mines like the great Sunnyside, the Silver Lake, and the Gold King were producing millions. Otto Mears' Rainbow Route Railroads were delivering concentrates to Silverton by the trainload, which were then shipped to the smelter in Durango on the Denver and Rio Grande Western Railroad. Silverton, the County Seat, was in her heyday, and during this decade saw water and sewer lines put in, a municipal electric light system installed and cement sidewalks poured. 1907 and 1908 also saw the construction of the new Silverton Town Hall, and up Greene Street, the San Juan County Courthouse. But, there was no good, modern hospital in this burgeoning little city.

The Union was at its peak membership in 1907 and was an integral part of the fabric of San Juan County's life. The Union sponsored dances, contests and horse races, and had a recreational program for boys. They had programs set up to help widows and orphans and the like, precursors to government programs which are common now, but were nonexistent then. At a vote of 1062 to 97, Union members voted to build a hospital for the use of its members and the general public. They voted to raise $35,000 dollars for the project by using existing funds in their treasury, holding benefits, and by assessing each member one dollar a month over and above his monthly dues of $1.50. In 1908 the firm of F.E. Edbrooke, Architects, of Denver was commissioned to draw up plans. Mr. Edbrooke was also the designer of the State Capitol, and is considered to be one of Colorado's preeminent architects. The amount of $1,200 dollars was spent acquiring the property at 13th and Snowden on which the hospital was to be built and A. Castonguay was hired as building contractor.

Edbrooke's plans called for a large building, 40 feet by 73 feet, to be built with specially hardened Denver brick, Durango sandstone, and with steps of Fort Collins stone. It was to be heated with steam and had a system of vents with electric suction fans to pull foul air out and pure air in. The basement was designed to have private rooms for the doctor and rooms for the laundry and boiler. The first floor was to have a large ward, six private rooms, a doctor's office, a medicine room, three restrooms and a waiting room. The second floor was to have another large ward, three bathrooms, seven private rooms and a large operating room. The operating room had large solarium windows and was to be lighted with the very best

The hospital cornerstone, Silverton Miner's Union Local #26 of the Western Federation of Mine, Mill, and Smelter Workers. *John Marshall Photo, 1994.*

lighting that electrical technology could provide, in case of nighttime use. The three floors were to be connected with an elevator.

The Union raised money from local Lodges, businessmen and saloon owners to furnish the hospital with the most modern medical tools. Its first doctor was A.L. Burnett. The hospital cost $28,000 dollars, and with the Union emblem on its fine cornerstone, served the citizens of San Juan County for the next thirty years. Its construction is a fine example of how the Miner's Union got involved in the community—almost every miner was a Union man, and events involving their families were frequent, including fund-raising for the hospital. Labor Day was a big holiday saluting working men and fought hard for by the Union. And much money was raised for the hospital in 1908 from the mining contests held in the middle of Greene Street, as well as three straight nights of dances with the 'best music.'

In 1916 Miner's Union Local #26 was taken over by the International Union of Mine, Mill and Smelter Workers, CIO.

How San Juan County acquired the Miner's Union Hospital and the Union Hall, now the theatre, is another story. The days of incredible wealth were now over, and the Depression had settled in. One of the few mines working, and by far, the largest employer in San Juan County, was the Shenandoah, managed by Charles Chase. Chase had been instrumental in introducing new methods of milling the base metals lead, zinc and copper for a better recovery, which allowed the mine and the Mayflower Mill to keep working while the mining industry was nearly shut down all over the rest of the West.

On November 4, 1938, Congress passed the Wages-Hours Law, or the Wagner Act. One provision of the law was that overtime wages must be paid after the first forty-four hours of work per week. The Shenandoah Mining Company countered by lowering the wage base rate for workers and requiring an eight hour work day. The Union was outraged and filed an unfair labor practices suit with the National Labor Relations Board. The Union had worked hard in earlier years to get the 'portal to portal' workday reduced to six hours, citing hazardous and hard working conditions, and had negotiated with mine owners for 'fair' wages. Shenandoah refused to budge, saying that its profit margin was so small that it would have to shut down without the decreased wage rate and increased working hours. On July 4, 1939, the Union voted 175 to 48 to strike.

At first the Union exhibited strong solidarity. The National headquarters set up a fund for the striking miners and sent A.S. Embree, a professional negotiator from Washing-

Beverly Rich, at home in the archives, usually knows of which she speaks. *John Marshall Photo.*

Gene Orr, Beverly's father, and Spider Stollsteimer are barring down at the Idarado, 1955. *Beverly Rich Photo*

ton D.C., to speak for the Union side. Other Locals from around the United States also contributed to the fund. Mr. Chase negotiated for Shenandoah. However, after a couple of weeks, talks broke down and Chase shut the mine down.

Times were hard and there was no work in the mining industry anywhere in the United States, in fact, there was no work to be had of any type, anywhere. After two months of picketing, the two sides were no closer than before and people started getting worried. Town and county fathers were concerned about the effect of the strike on the tax base and businessmen were concerned about the effects that the strike was having on the local economy. *The Silverton Standard and the Miner* called for an end to the strike and called A.S. Embree an 'outside agitator.' The mood in town was tense.

On August 28th a meeting was held at the Miner's Union Hall. There were at least a thousand people outside of the Hall, many of them wives of members and nonmembers. According to later court proceedings, several members tried to speak but were not recognized by the officers of the Union, but instead were fined. The officers said they were trying to subvert parliamentary procedure and that several of them were not in 'good standing' as members.

The Miner's Union Hospital building in 1994, awaiting restoration and a new use. *John Marshall Photo.*

A fight ensued, and the Sheriff and the Chief of Police were called in. They advised Embree and the officers of the Union to leave town for their own safety and escorted them out. After they left, at one o'clock in the morning, new officers were elected, and the International Union of Mine, Mill and Smelter Workers, CIO, Local #26 was dissolved, and a new union, the San Juan Federation of Mine, Mill and Smelter Workers was formed. All of the old Union's assets were turned over to the new Union, including the Hall and the Hospital.

By this time, Local #26's membership had been reduced to just a few people, most former members having joined the new Union or left town in the wake of the bitterness of the strike. The old Union sued for the return of the assets and for $50,000 dollars in damages stemming from beatings and assault. They got the property, but only for a while.

In the fall of 1938, the Colorado Supreme Court rendered a decision that all property of the Miner's Union #26 was subject to taxation. In a resolution on September 5th, 1939, the County Commissioners ordered that the property be put on the tax roll and be taxed for *all* of the years that it had been omitted—over thirty years. The Union then sued the County Treasurer, saying that the hospital was used for charitable purposes and should be tax exempt. The Union did not pay the taxes and the County assumed ownership of the hospital and the Union Hall on a Treasurer's Deed. The suit was settled out of court ten years later in 1948, but only under these terms: that the Union would dismiss with prejudice the suit against the County Treasurer and in return the County would pay the Union $4,000 dollars for all of its property in Silverton. It was the advice of the County's attorneys that "there were many uncertainties connected with the claims of the County for delinquent taxes, which would probably mean extensive and prolonged litigation at considerable cost to the County."

In the years since then, San Juan County sold the Union Hall to American Legion post #14 which still owns it today. The hospital ran until the mid-1950's, and then it was used as a clinic until the early 1970's until the Carriage House clinic was built. It has since been used as an office building, and San Juan County is looking for ways to rehabilitate the building, perhaps as classroom space for some colleges that have expressed an interest.

And because people cared, new people would be drawn here. Through their energy and efforts intertwined with those already here, the town would grow again and even prosper. ❋

—From Beverly Rich, San Juan County Treasurer and Chairperson of the San Juan County Historical Society.

John Marshall Photo

Christ, of the Mines

Yet, prosperity for any mining town is a cyclical condition. A mine can loose its backing. A mine can run out of ore. In 1952 the 44 raise of the Shenandoah-Dives collapsed. What remained was an impenetrable pile of timber and debris. The main access to the upper workings had been lost. Fortunately, no one was hurt in the collapse, but many would be affected by this event which eventually led to the closing of the mine. A grant was obtained from the government to further explore the main level areas. When these efforts proved fruitless the mine was closed.

Hard times besieged the town of Silverton. For the first time in as long as many could remember, there wasn't a working mine in San Juan County. In January of 1958 the Catholic Men's Club conceived the idea of a shrine, and with the guidance of Father Halloran of St. Patrick's Church, the site on Anvil Mountain overlooking town was prepared. Land was donated, local materials were used (the stone around the statue came from the old Fisher Brewery), and labor came from city and county crews as well as from many local people. The necessary finances were raised and by 1959 the Sacred Heart of Jesus, made of carrara marble carved in Italy, was guarding and protecting the town.

Shortly after, Standard Uranium appeared, driving the American Tunnel (while at the same time spending two years at the Shenandoah) to reopen what had once been Colorado's largest gold mine, the Sunnyside. Because of the American Tunnel, the lower extensions of the Sunnyside veins became accessible, and mining, this time from the old ghost town of Gladstone once more became the mainstay of Silverton and San Juan County. Mining was back. If a statue brought the town good luck, so be it. The town would stay. Besides mining, other means of reaching financial strength would be explored in this enduring town.

'Butch' Salfisberg Dies Of Gun-shot Wound

Tragedy struck the Silverton community and everyone was shocked and saddened when young "Butch" Salfisberg passed away at the San Juan Hospital early Thursday morning less than twelve hours after he had been shot by a 22 caliber rifle. Elario Posthain, about 65 years of age, is being held in the Silverton City Jail, charged with discharging the fatal shot that took the life of this 9-year-old Silverton youngster.

The shooting took place around 4 o'clock Wednesday afternoon as a number of Silverton school children were playing on the hill immediately back of Posthain's cabin, which is but a short distance from the school building. It is alleged that Posthain came out of his back door armed with the rifle as Butch Salfisberg and Irvin Blackmore were playing on the trail above. He admitted pointing the gun toward the boys, shooting, then returning to the house. Butch was struck on the right side, just below the ribs. The bullet ranged up, piercing both lungs and lodging in the back of the left shoulder. After being shot Butch and Irvin ran to the bottom of the trail at which point Butch collapsed. He was taken immediately to the hospital which was only one block distance. There Dr. Holt and his staff of nurses went to work to save the boy's life. He passed away about 3 o'clock the next morning.

Ross Beaber, editor of the Silverton Standard and the Miner *holding an original Polariod, 1958.* *Paul Beaber Collection*

A Silverton Scrapbook
Snapshots of Life in Town
Changing Faces, Changing Places, Changing Times

There were movie theatres for entertainment and, thanks to people like Ross Beaber, there were newspapers to be read. Ross moved to Silverton with his wife Francis in 1940 and took over the newspaper. Years later one of his sons Larry wrote about that purchase: "It would have discouraged most men—what he faced in Silverton in the spring of 1940: a large portion of ill will left over in the townsfolk against the newspaper from a bitter union fight; a small and delinquent subscription list; a lot of old, antiquated printing equipment—that was about it." Yet, until January of 1963 Ross kept the paper coming, week in and week out. And the next owner? A gentleman by the name of Allen Nossaman, now a County Judge and Silverton historian *par excellence.*

With Ross writing in his paper about the golden future of Silverton, the train from Durango, instead of quitting as less ore was shipped over its tracks, turned to a new source of revenue—the tourist. Under Ross's optimistic writings, the roads into town gradually improved, and another option opened for those intrepid tourist souls. Now in Silverton many thousands come every day over the roads, over the rails, to enjoy our little mountain town. However, a newspaper covers more than tourism. Necessarily,not all Ross wrote about was good news.

Francis Adair Salfisberg, known by most everyone here as "Butch," was born in Silverton on January 14, 1942. He was in his third year of school. He was an intelligent and lovable youngster and popular with his schoolmates. For a young chap, he was noted for his ruggedness and stamina and was ready to keep going while others were fatigued by his over-supply of energy. He was a good friend of the Standard publisher and his family and no child will be more missed about us, except it be our own.

Besides his parents, Mr. and Mrs. Frank Salisberg Butch is survived by two sisters, Roberta and Mildred and one brother, Danny; a grandmother, Mrs. Richard Cram of Durango, two aunts, Mrs. Agatha Peterson of Silverton, Mrs. Alice Kornafel of Denver; two uncles, Fred Salfisberg of Cheyenne, Wyoming and Arthur Berkey of Los Angles; also several cousins and many good friends.

Funeral services will be held at the Congregational Church Saturday afternoon at 2:00 o'clock with Rev. Burke in charge. The Maguire Chapel will direct the service and burial will be in Silverton's Hillside cemetery.

District Attorney George Dilts of Cortez and Assist. District Attorney Byron Bradford of Durango were in Silverton most of the day Thursday gathering evidence on the shooting. Coroner Maguire impaneled a jury consisting of Robert Wyman, Herman Zueck, Alcide Fedrizzi, Ragner Beckman, Auborn Ramsey, James Drobnick, and John Troglia. The district attorneys said charges would probably be filed within the next few days.

Butch Salfisberg gives Rory Calhoun a few roping tips for the movie Ticket to Tomahawk, *being filmed beside Molas Lake in 1949.* *Dan Salfisberg Collection*

So often for the residents of this mountain community, life was too short. This sign was placed by the Lime Creek Road turnoff on the south side of Molas Pass. *Paul Beaber Collection*

Several movies were made in Silverton. This train scene is from the movie **Ticket to Tomahawk** *which would play as the last performance of the Lode Theatre on January 27, 1990. No theater has operated since.*

Nomad on the Tracks

The Nomad, original private car of the late and prominent first president of the D&RGWRR, General Jackson Palmer, made its reappearance this week as part of the Silverton train that came into Silverton Sunday. The beautiful old deluxe car has been completely restored by the owners, H. B. Wood and Ted M. White, prominent men of the San Juan basin.

The car will make occasional trips this summer through special arrangements by the owners and the railroad. This historically famous car's interior is of mahogony. The furniture is wicker with red plush seats. Red velvet draperies and real tassels are in taste. There is an observation platform, gas lights, kitchen complete with refrigerator. It will comfortably house seven, including sleeping quarters.

Mr. William Riggs, manager of the Grand Imperial Hotel, Mrs. Pauline Hopkins, Mrs. Martha Binford, Mrs. Louis Sparkman, and Mr. Gideon Binford were guests of Mr. and Mrs. H. B. Wood on Sunday when the party made the maiden run on the private railroad car, the Nomad. Others in the party included Colonel John Whitfield, re- U. S. Army, Mr. Kenneth R. Menk of Durango and Mr. Wood's house guest, Mr. Fritz Klinke from Palo Alto, California, and Mr. Wood's son, Mark. Mr. Riggs and his party were met in Durango by the Hotel car and returned to Silverton, bringing Mr. Klinke back with them for a visit to Silverton.

The last picture show!

Ticket to Tomahawk

In memory of The Lode and Star Theatres, 1916 - 1990

Benefit San Juan County Historical Society

January 27, 1990 - Admission $5

For a long time there were two movie theaters in town. One was the Star Theatre. Helen Crawford, in 1906, at the age of eleven, learned to play the piano. How? By reading lessons printed in the Denver Post. Besides playing at many social events, she played for years at the Star, accompanying such silent greats as Douglas Fairbanks and Mary Pickford. Louis Wyman wrote a memorial for her in 1972. One of her brothers, a Bausman, prospected the mine pictured on the back cover of this book.

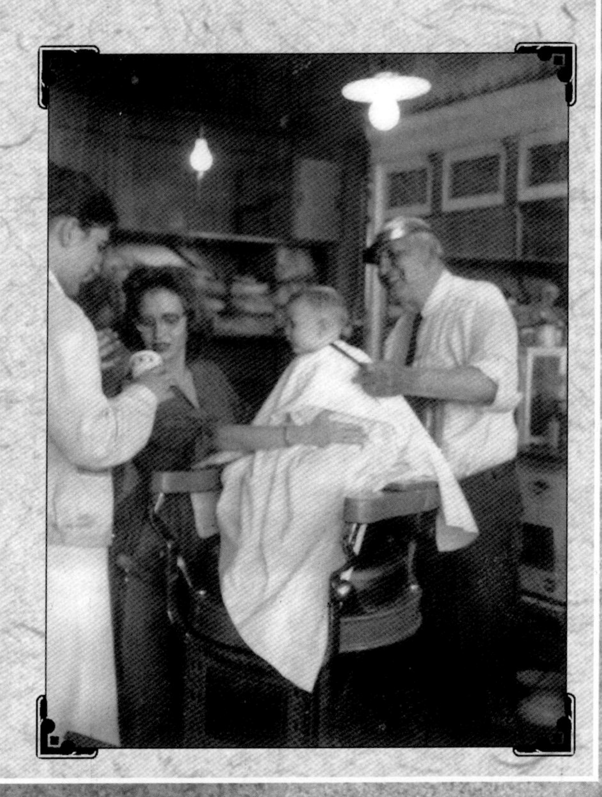

Other children were luckier and were able to grow up. The first haircut. It's a big deal. Here, George Bolen, the barber, gets ready to cut his great grandson's hair. Jonny Rae, the victim, is told not to squirm. Lois Cole (Jim Cole's sister and granddaughter of George), holds Jon while Diane Girodo helps out. *Jim Cole Collection*

The train brought people who were thinking of the future. Fourth from the left rides a young Fritz Klinke (holding a Coke). It's the first trip since the forties for the deluxe rail car, the Nomad (still available today). The year is 1957 and the car is northbound, stopped at Tank Creek. Fritz would return permanently thirteen years later. *Fritz Klinke Collection*

Fritz and Jan Klinke stand proudly in their new restaurant, The Pickle Barrel, in their first summer in Silverton, 1971. The restaurant remains under the same ownership today, twenty-five years later.
Karl Zimmerman Photo

Blair Street, Silverton, between 11th and 12th Streets. From right to left is a white boarding house which burned and later was rebuilt as the Ore Bucket. A vacant lot lies between that and the Zanoni Padroni Bar which later became the Bent Elbow. Next door to the bar was Big Tilley's Tremont Saloon (with second-story porch) which later became the first Bent Elbow until it burned in January of 1968. The big building on the far left was a working girls building known as the Green House and today as the Shady Lady. The two small buildings are former cribs and the one to the right was the office and living quarters of Doctor Holt. The building was bulldozed in order to put out the smoldering wreckage of the Bent Elbow fire. *Tom Savich Collection, 1958*

(below) The first Bent Elbow that burned. In the back was an addition that connected to the building housing the Bent Elbow today. That building was called the Monte Carlo Mercantile and was owned by Effie Andreatta and rented to Thelma Zanoni who ran a candy store and gift shop.

Pictured here is the owner Frank Bostick, wearing a holster. Frank was the creator of the original Bent Elbow. He was also a prime developer of Blair Street and with the help of Tom Savich hauled old buildings down from Middleton for Old Town Square across the street. The saucy women seated are Martha Jane Oliver and Frank's wife, Carolyn Bostick, on the right. First at the bar is Ray Irwin. He looks toward the large painting on the left-hand wall which survived the fire somehow to be hung again in the new Bent Elbow. The melodrama players pictured here were directed by a youthful Bill Maguire. Two of the other actors were Jerry and Tommy Sandell.
Tom Savich Collection circa 1960

A man of many careers, Herold Waddington appeared in Silverton in 1940. This talented gentleman turned his attention in 1953 to the proposal by the Silverton narrow gauge, known as the Denver and Rio Grande Western, to abandon its line from Durango to Silverton. He organized the Animas Canyon Railroad Promotion Committee, tirelessly campaigning with efforts that stretched across the country. Abandonment? Request denied. By 1963, 50,000 people had travelled those rails and today that number is over 200,000 a year.

Chris Nelson Collection, 1958

(above) R.M. 'Snarkey' Andreatta and his wife, Effie Weaver Andreatta bought the Bent Elbow and then created the new one next door after the fire. Meals are still served daily. The former Highland Mary mining superintendent operated the restaurant until his death in 1972. Then Effie ran it with the help of her son Mike until her death in 1995. *Mike Andreatta Collection*

(left) Today, Mike runs the restaurant, still a family affair. From 1962 until 1991 he also put in his time mining with Standard Metals, and was never against picking up a part time job along the way. Here Mike works on the restoration of City Hall after it burned in 1992. *Zeke Zanoni Photo*

Born in 1902, 'Noni'—Mary Dallavalle Swanson—saw 87 years in Silverton and all that went with it.

Gerald Swanson Collection

"When you served the public you could never tell who might come through your door. Some days, it was good just to have a customer, any customer."

Weathering the Changes

Mary's father, John Dalla, had died in 1911 from frostbite suffered while driving cattle over Molas Pass in a terrible winter storm. Mary's brother Charles had died in 1914. Her mother, Domenica, passed away, along with two of Domenica's sons, John and Beno, in the flu epidemic of 1918. Mary took over the operation of the family boarding house and raised her remaining five brothers and two sisters.

Time passed to the year 1937. The Antonelli family wanted to trade a building they owned for Mary's boarding house. That building stood on Blair Street and had been built in 1901. The stone had been quarried in Bertramsville and was brought over by the railroad. (In 1989 Mary was interviewed by Steve Logan, Silverton radio announcer, just shortly before her death, and related something about her accomplishments.) "I was afraid I was going to lose that store for back taxes. Money was scarce in 1937. I asked Centennial Insurance in Silverton for a loan to pay the taxes. Amazingly they gave it to me and eventually I paid it off at five dollars a month."

In 1941 Mary started Swanson's Market with $25 dollars. Charlie Chase was running the Shenandoah-Dives that year, and it was the only mine working in the county. As World War II progressed, more mines opened and Mary's new business began to prosper. Mary's entrepreneurship, with the help of her son Gerald, kept the store open throughout the years. In 1974 she was given the Small Business Award for the district including Silverton for her success as a woman running a small business. To Mary, that was as good as being Miss America. She was elated. Retirement finally came for her in 1986 at the age of eighty-four after 45 years in business.

Gerald Swanson, son of Mary Swanson, stationed behind the meat counter he operated for 37 years (1955 to 1991 until the market closed). He was the last-full time butcher in Silverton. Prices look pretty good! Easter, 1982.

Gerald Swanson Collection

Mary Swanson with some of her boyfriends in front of the Dalla Boarding House which she ran after her mother died. (Well, they might have been boarders.) Gerald Swanson Collection

"In 1920 we rode the train, all of us girls, from Silverton, up to Eureka to the New Year's Eve Dance. One year we had the chance to go to the Sunnyside Mine for a party and dinner at one o'clock in the morning. We could ride the tram or take a buggy. I took the buggy. I was too scared of the tram. But, oh, we had fun. I loved to dance. What kind of dance? Oh, waltzes, the two-step, the foxtrot. No, I never did learn to jiggle. You know, like they do today."
—Mary Swanson

The timeless joy and sadness and beauty of life...

(left) Charles Dalla, 1910 to 1914. He died of burns suffered from a fall into boiling water two weeks before his fourth birthday.

(right) Angelo Dalla, brother of Mary Swanson, at age twenty. He sits next to the boarding house just after his mother's funeral. She had died in the flu along with two of Angelo's brothers in 1918. Gerald Swanson Collection

SHE DIED A BEAUTY OF REPUTE

HER OTHER VIRTUES IN DISPUTE

N. Scott Momaday on The Death of Beauty

It was a Fourth of July parade in 1976 and Gerald Swanson was proudly advertising his meat market, or was it dairy products? ***Gerald Swanson Collection***

Marion Palovich has his picture taken with Tom and Margie Savich, around 1928. Marion would procure business for Jew Fanny, one of the working girls. Tom Savich Collection

"Jew Fanny? Sure, I knew her. She was the last lady on the line. She used to come into the store all the time. We started the store in 1941. Always she'd be smoking. She'd walk around the store puffing on a cigarette, blowing smoke out like this. 'Phew! Fanny,' I'd say, 'why are you smoking all the time?' 'Oh, I just vant to be sociable,' she'd say. She was always dressed up nice with a hat and stuff. Later on when she was moving (to Denver I think) I asked her, 'Why are you moving?' 'Well Mary, it's gotten to the point a decent woman can't make a living in Silverton anymore. There are too many women giving it away!'"

—Mary Swanson, from Allan Bird recording and Gerald Swanson.

(left) Ralph Weaver and his daughter on the way to Swanson's Market, 1957.

(right) Now that's some bull. The rest of the story about this ride of Gerald's is too. ***Gerald Swanson Collection***

Floor Lifting?

Roof Raising?

Better Let George Do It

"Let George do it." It is an old saying and it is usually uttered by people too lazy to do a particular thing on their own.

But when it comes to raising the floor of a 97-foot-long building two feet higher than it has been since the turn of the century, you had better let George Bingel do it.

Bingel, operator of the Eight Ball, a successful recreation center for Silverton's young people, decided that a nice touch would be addition of a two-lane bowling alley in the basement of the center.

The big problem Bingel faced, however, was that the basement was about two feet shy on headroom. Normally that would have meant breaking up the existing concrete floor, digging the basement two feet deeper and then pouring a new floor -- no small project. Normally.

Bingel saw it differently. Since headroom was the problem, why not jack up the main floor (the basement's ceiling)? That solution not only would increase the headroom in the basement, but would decrease it upstairs where the ceiling was very high.

A number of technical problems had to be solved in order to accomplish this feat. Bingel had to devise a quick way of grabbing the major supporting posts in the basement for the jacking process. He had to remove the top portion of a partition upstairs. He had to insure the floor would rise free of the walls of the building as he jacked. And he had to engineer a way to jack up the floor so it would again be level. Bingel said he spent four or five days getting ready.

He then hung a "closed for alterations" sign in the window of the Eight Ball and, in three days, raised the roof -- or the floor, depending on your viewpoint. The jacking was all done by Bingel using a mechanical "track jack." Boosting each post up in four-inch increments, he avoided damage to the flooring upstairs. Using an engineer's optical level, he was able to check out his progress and perform final leveling operations without leaving the basement.

Bingel says that he is shooting for an opening day sometime in December for the bowling alley. He says there is still a lot of work to be done in the basement before the alley can be operated. Some of that work might be in getting near 100-foot bowling lanes into the basement, but if you are wondering if it can be done, well, let's let George do it.

We stopped George Bingel from working in the basement of his recreation center long enough to get an idea of how he got his floor to rise from the place it was (where his left hand is) to where it is now (two feet higher). He worked alone and had the hardest part of the job done in three days. The basement is slated to house two bowling lanes.

Photo, article and caption from* The Silverton Standard, *October 27, 1972.

If you've got a growing town, maybe you should have a bowling alley. Roof's not high enough? Heck, no problem, get an old miner. Let George do it!

Buildings Changing

George almost talked Wilma into a little para-sailing on Lake Havasu. No luck there, so he went alone—just six months before his death at age 73. ***Wilma Bingel Photo, 1990***

Versatility

Libraries and theatre, in varying degrees of artistry, seem natural companions.

The Carnegie Library, built of brick in 1906 with $12,000 dollars of Carnegie-funded money. *John Marshall Photo, 1994*

Gerald Swanson shone as the Inspector in **Carousel,** ***presented at the Grand Imperial Hotel in 1977.*** *Gerald Swanson Collection*

Bill Maguire in the nineties. A graduate of the University of Northern Colorado in Performing Arts, he directed melodramas in town in the fifties in Frank Bostock's Bent Elbow. Bill found out fairly quickly other ways to make better money in Silverton—mining, milling, trucking and even county commissioner became his pursuits. *Judy Zimmerman Photo*

This might be as close as some of these men get to the library (just don't bet on it).

(left to right, bottom to top in photo above) Andy Hanahan, Vernon Ince, John Wright, Montana Mike, Keith Meader, Alan Wilke and Mike Cummins. They are all gathered here for Andy's fiftieth birthday party. It was a special occasion to break a 'curse' from Ireland, whereby nobody in his family had yet lived to be fifty. Keith is resting in the casket that was supposed to have been for Andy's funeral. It must have worked for one can still find Andy around town seventeen years later. ***Andy Hanahan Collection, 1978***

Upstairs in the Carnegie Library. Downstairs had been a men's club and reading room. It has since been restored by the Library Board and is now used by all the residents. ***John Marshall Photo, 1994***

John Ross, a wonderful storyteller and talented man, played at the Beaumont Theatre in Ouray in 1964. The melodrama travelled to Lake City and, of course, Silverton. John bought the San Juan Cafe and continued to produce melodramas there. ***Andy Hanahan Collection***

George Crane started a one-man woodshop in a house on Reese Street in 1976, fulfilling his dream to be in the mountains, climbing and skiing, and, oh yes, running a woodshop. Things were fine for a while, but by 1987 he had four employees and felt the pull of the bigger market of Durango. Today, it's the same business but with several delivery trucks a day and forty employees. The mountains are on hold. This is George in his shop in 1978. *George Crane Collection*

Tim Lane, officiating at Mark and Dotty Beaudin's wedding, 1985. *Mark Beaudin Collection*

Different people would arrive in town for different reasons.

Flexibility

Changing hats often allowed people to pursue a variety of fascinating careers. Right here. And that allowed for a lot of interesting and talented people to gather here. In 1972 noise was coming from our neighbor to the north, Ouray. It was felt by some that cloud seeding projects were causing increased avalanches and their inherent dangers in the San Juans. Ed LaChapelle from Alta, Utah, was awarded a contract from the Bureau of Reclamation to investigate. He came with his wife, Dolores, and hired Dick Armstrong to head the project. Many knowledgeable people became involved in this snow study over the years, temporarily making Silverton their base.

Around 1979, Tim Lane and Jerry Roberts became involved in the ongoing study. With money more scarce than housing, they ended up living in tents inside a partially burned-out house on Main Street, known to a few as the Charcoal Palace. Eventually two cabins in Chattanooga, seven miles north of town, became available. There was even glass in the windows. Soon parts of the back country that hadn't seen people for many years were finding ski tracks over their quiet winter coats. Tim finally ended up in Portillo, Chile, running an avalanche program for the largest ski area in the southern hemisphere. Jerry migrated north, stopping just past Ridgway, and is still climbing, skiing and guiding.

Jerry Roberts, Tim Lane's partner in crime, warming up, waiting for an engine. *Jerry Roberts Collection*

Though she's been gone from town for sometime now, there are more than a handful of stories still floating around about this hell-raising, loving, ornery, ready to fight, cuss or curl hair woman. That's right. Lucille's Beauty Shop in front and the Silverton Dude Corral in back. Lucille Bowman came to Silverton in 1927 when she was only fifteen years old. She stayed until 1983 when she went to Grand Junction to live out her remaining four years of life. Her former customers recalled being left under the dryer until things got too hot, only to find her out back chatting up a new cowboy by the corral—a corral she ran for twenty-five years. She was the first woman in the state to hold an outfitter's license. Things were never dull around Lucille. This picture was made possible by Jack Caine's son and family in Grand Junction. Jack Caine, Senior, had been married to Lucille until his early death in 1962. Jack was the Sheriff of San Juan County at the time.

Virgil Mason was sheriff of San Juan County for ten years. Virgil died in 1978 and his headstone has his Sheriff's badge and picture on it. The photo at right appeared on the campaign poster for Virgil's third term. Don Bachman (plaid shirt) would head the Colorado Department of Transportation Avalanche Forecast Center in Silverton in the winter of 1992-93. Search and Rescue personnel gathered in support of the Sheriff in the Grand Imperial lobby in 1977. Andy Hanahan Collection

Whitey David is seen here on the deck at Bobby Blower's wedding in 1984. Two days later he was dead in a heavy equipment accident up Cunningham Gulch. Whitey spent most of his time in various mining adventures around Silverton from 1941 until his death. Jean David Collection

Sheriff Mason
for
SHERIFF

He knows the job
He knows the people

Virgil F. Mason Sr.
Republican candidate for San Juan County Sheriff
Courage, Integrity & Compassion

July 4th Celebrations and Fire Department Centennial, 1979

Kneeling in front, left to right: *Zeke Zanoni, Louis Girodo, Lorenzo Groff, Andy Archuleta, John Archuleta, Tom Zanoni*

At left of wheel, left to right: *Jim Hook, Lloyd Rogers, Gwen Taylor, Robert Baca, John Jacobs, Bill Gardner, Charles Moore, Hal Slade*

At right of wheel, left to right: *Fred Wolfe, Ernie Kuhlman*

Band, left to right: *Richard Bovee, Terry Morris, Dale Meyers, Todd Chapman, Kathy Lund, Scott Smith, Allen Nossaman, Janet Sharp, Roy Perino, Gerald Swanson, Gary Miller (kneeling)* **At left corner of building:** *Scott Wallace*

At left and on top of fire truck, left to right: *Jim Alabashi, Don Gurule, Gilbert Archuleta, Brant Moiseve, Tom Hart*

Moms from the largest LaMaze class ever held in Silverton, 1982.

Front row, left to right: Ruth Ward (instructor) holding twin Kristin Carlson, Vicki Willis Morris with Michael, Karin Johnston with Sidney, Pam Pekrul with Gaelan, Robin Dunn with daughter.

Back row, left to right: Carol Carlson with Jordan, Mary Spense with Luke, Cindy MacDougall with Malcom, Susan Hembree with Paul, Judy Roberson with Emily. *Ruth Ward Collection*

New Life

Fittingly, after so many births that year, the theme for the Fourth of July in 1982 was **babies!** *A few of our old friends led the parade dressed in diapers. Their bottles weren't milk however.*

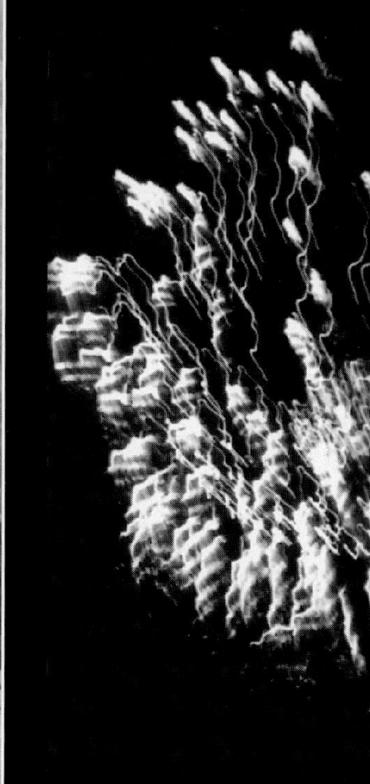

What? You never knew there were hot air balloons in Silverton? Neither did I. Hot air? Yes. Balloons, no. Hardrockers Holidays, 1978. This celebration may be all that's left of mining in the San Juan. ***Gerald Swanson Photo***

The Carriage House is under construction in 1981 and when built it housed the Fire Department, the Ambulance Association, a clinic, Search and Rescue, the Public Works office and their equipment. It was a key building for the town, built with a five percent FHA loan, one of the few given that year in the state of Colorado. ***Gerald Swanson Photo***

Otto Smith in front of the Benson building with Cindy Luther and his daughter Tiffany in 1976. Otto was the son of Annie Smith and grandson of Frank Anesi, an early Silverton resident who came to town in 1901. It was his grandfather's building, the Benson, that Otto purchased and restored, turning it into gift shops with apartments upstairs. ***Photo courtesy of Tiffany Smith DeKay***

Hardrockers fill 13th and Greene Street around 1950. ***Paul Beaber Collection.***

Tourism. Town. Mining. Each would grow or even die in fits and spurts, sometimes logical and ordered, more often not.

The mining celebration, Hardrockers Holidays, would flourish through the years, stumbling here and there, but still hanging on today, just barely.

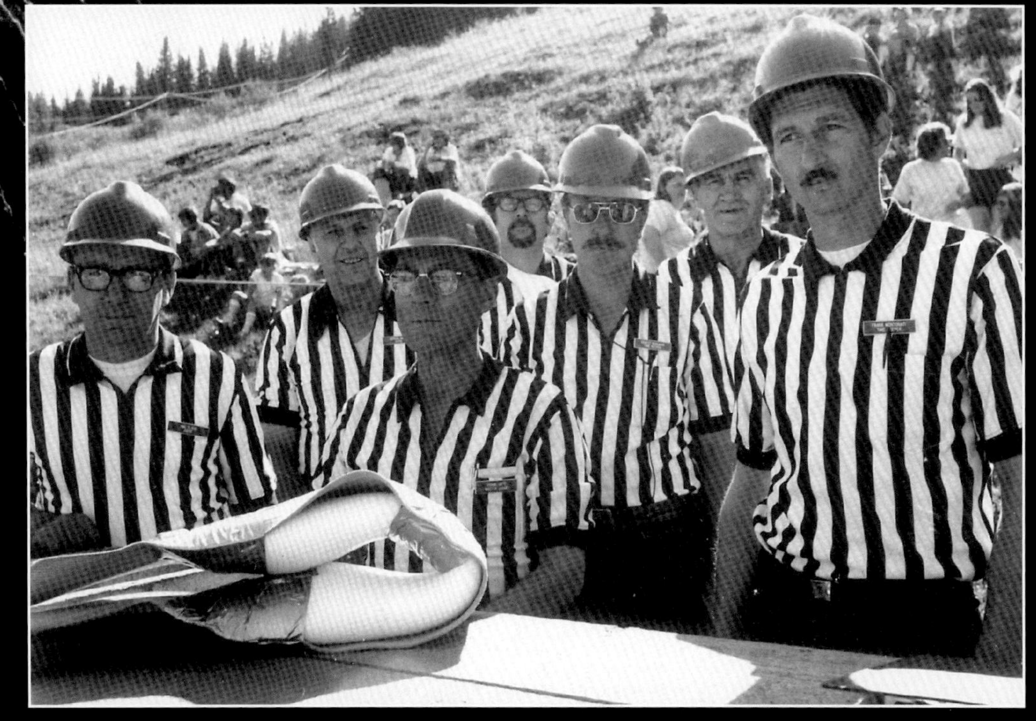

The judges at Hardrockers, 1977. (left to right) Hal Slade, Herman Dalla, Master Judge Frank Hitti, Gerald Swanson, Ernie Kuhlman, Bob Ward, Frank Montonati. A rather stern group. ***Gerald Swanson Collection.***

Some men would work underground, then above ground, then head back underground. They stayed here, through thick and thin, doing whatever it took, mining was in their blood. Jim Case on Treasure Mountain doing exploratory core drilling in 1978. **Dale Butts Photo**

In the sixties and seventies and on into the eighties there were a lot of small operations that started on high hopes and stopped when economic reality set in: Pride of the West, Baker's Park, The Gold King, Yukon Mining and Milling, Buffalo Boy, Henrietta, Gary Owen, Montana, Old Hundred, Bausman Mine, Black Prince, Anti-Periodic, Big Colorado, Intersection, Belle Creole, Hercules, Prodigal Son, Pride of the Alps, Lackawanna, Klondyke, Scotia, Silverton Girl, Zuni, Vermillion, Lead Carbonate, Galty Boy, Hoosier Boy, Golden Fleece, Senior Warden...

Mines and Miners Old and New

(above) Some folks came here with different talents, to enjoy the mountains as much as anything while mining usually provided the work. When the mines played out, they'd move on. Tom Tisdale, a talented musician, sits on top of Bear Peak. He worked at the Ezra R. and Standard Metals before it became time to go. **Gary Leidig Photo, 4th of July 1981**

(left) The old Lackawanna Mill, on the edge of town, as it appeared in 1974. Today, the structure endures, an occasional board banging idly in the breeze. **Gene Harrison Photo, Gerald Swanson Collection**

David Tisdale worked the mines along with his brother Tom. For a while they lived in a teepee only a mountain goat could find. Then they were gone, 1980. ***Gary Leidig Photo***

There was life up Arrastra again. The Ezra R. and the Grey Eagle would see life for a while. And down by the Mayflower Mill, the Valley Forge would be worked again. These guys were the crew.

Some had worked here before, they came back. Others had mined elsewhere, some were new and had a lot to learn. They all worked hard, and when the jobs ran out, most would move on. The houses they had bought would be sold and their dreams were forced to change.

(above) Fred Chase and Allen Wallace in the Grey Eagle. It was a four-foot high drift for one hundred fifty feet. A true "Cousin Jack" drift. ***Gary Leidig Collection***

(above right) Tom Zanoni came in and added direction to the group with his years of mining experience. ***Gary Leidig Collection***

(right) Mark Mohnac found that the light at the end of the tunnel was a degree in chiropractic medicine. ***John Marshall Photo***

Gary Leidig in the doghouse of the Ezra R., 1980. He became a cowboy in New Mexico. ***Mark Mohnac Photo***

Amos Jaramillo at the Idarado. Mining was more than a job, it was a way of life. This picture was engraved on his tombstone when he died. He worked at the Shenandoah from 1947 to 1953 and then it was off to the Idarado for twenty-five years until the mine closed in 1978. His daughter, Christine Bass, still lives in Silverton, helping others as the county's Director of Emergency Medical Services. *Christine Bass Family Collection*

THE IDARADO was just out of San Juan County on the other side of Red Mountain. It was a good mine, good pay and well run. A lot of men from Silverton worked there, driving over the pass each day or riding the company bus, back and forth, all hours of the day and night.

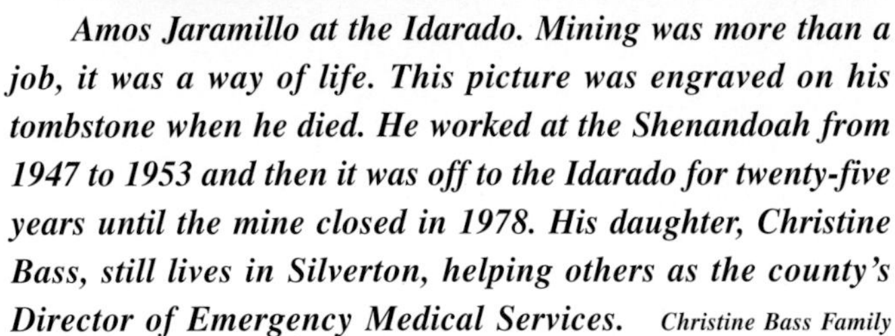

The above-ground buildings of the Idarado as it sat on the north side of Red Mountain around 1975.

Zeke Zanoni Collection

The innocence of youth allowed me to grow up with a fearless admiration of mining and the men who spent their lives working deep inside the dark damp earth. With age and exposure to mining life my opinion of a miner's life changed.

Father was a Miner

I can remember being a small child and hearing the alarm clock ringing in the very early hours of morning, followed by my parents' muffled movements in the kitchen. My mother would get up early along with my father to start the coffee, pack his lunch, and have those few special moments with my father before he left the house, headed for the bus that carried the miners to work. It never crossed my mind that by the time my day was just beginning, my father was deep in the mine, working to provide for my family. When my father returned home for the day, Mom would have dinner ready and waiting. My father would clean up and our family would have dinner together.

Sometimes, after coming home, my father would quietly mention there was something in his lunch box for us. I never realized the link my father's lunch box was between us, until years after he no longer needed it. As little children, my brothers and sisters and I used to love to walk the two blocks to Main Street to wait for the mine bus. Mother wouldn't allow us to cross Main Street so we'd stand on the corner by the drug store and wait for my Dad to cross over to us. He'd hand his lunch box to one of us and gather the other siblings by the hand, and we'd all walk home together. Dad always saved us a cookie or two, or a few slices of an orange or whatever goodie Mom had packed for him. We'd open the lunch box, the familiar smells from the mine hitting our noses, and the mine dirt getting on our hands. The goodies still tasted wonderful to us.

As we got older the goodies were replaced with crystals and mineral specimens. Dad would carefully bring them out and tell us where he'd found them and explain the different properties of the minerals. It always amazed me after they had been washed, how they sparkled and shone. Mom would pick out the ones she wanted for her personal collection and she would let us kids have what was left. At that time they were considered waste products and all the men brought them home for their kids to sell to the tourists in the summer. I can remember feeling great satisfaction in being able to rattle off the different mineral names and tell the tourists about 'fool's gold.'

By the time I was eleven or twelve I started helping clean my Dad's lunch box and sometimes packing it to help out my mother. I could often tell what kind of day my father had by the food he didn't eat or the condition of the outside of the box. I was beginning to realize more what my father, an older brother and two cousins who worked in the mine as well, went into when they went to work.

Reality was driven completely home when Marge Baudino knocked on the door in the middle of the day to tell us my brother, Chris, had been involved in an accident at work that day. Only that morning my brother, who had a cold, had stated, "If I could just stay home and go to bed today I think I might live." But he was saving to buy a truck and didn't want to miss work. "I'm sorry" is such a hard thing to hear and to have to say.

Christine Bass, daughter of Amos and May Jaramillo, outside her home in Silverton, 1994.

John Marshall Photo

A few years later my brother-in-law joined Chris in being a mining statistic, leaving my expectant sister with a son who was never to know his father. I was married to a miner then and the youthful innocence no longer protected me from the ever-present, though hidden fear, of hoping no one I knew or lived with would be hurt or killed. Each day started with a silent prayer to keep the men safe.

My father retired in 1978, after working in the mines for the greater part of his life. I watched him fight a losing battle with silicosis for the next three years and did what I could to make his remaining days comfortable. A casualty of mining, he passed away on a September day in 1981.

Working as a medic I cared for several injured miners. My ambulance team would be called to the mine or the mill. We never knew until we got there how badly broken or bleeding our patient would be. We'd arrive to find coworkers helping to move their partner from the mantrip out to us. He would be bound in a litter, trying to be strong, waiting for someone to take away the pain. In trying to learn what had happened to my patient, I was always struck by the camaraderie and worry I could read in the miner's eyes as they hoped for the best for their friend. I knew they were thinking but for the grace of God it could be them. They were all subject to the same dangers.

After living my life around mining and miners, listening to stories and watching many of them played out in the lives of friends and loved ones, I finally have a true respect for my father and what he did to make a living for us. While there is a certain sadness at mining becoming a thing of the past, I am glad my son is not a miner of the future. ❋

(left) Joe E. Salazar was at the Idarado, Camp Bird, Little Dora and then twenty years at the Mayflower Mill where he died on the job in 1972 at fifty-five years old. *Mannie Salazar Collection*

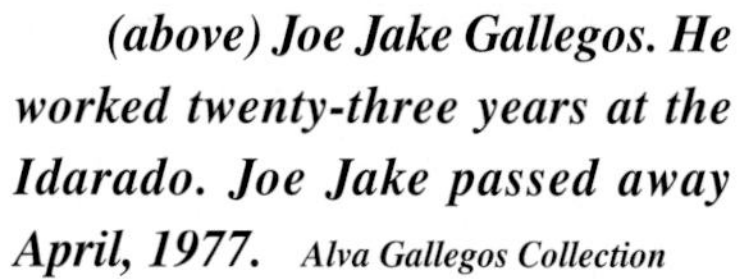

(above) Joe Jake Gallegos. He worked twenty-three years at the Idarado. Joe Jake passed away April, 1977. *Alva Gallegos Collection*

(below) Demecio 'Speedy' Gallegos put in his time at the Idarado and Standard Metals. *John Marshall Photo*

The men would go into the mines and mills facing the daily dangers of production, yet, it was the women who would hold the families together, often even working jobs themselves. Each day they faced the fear that their men might not come home. Still, they persevered.

(center) It would be difficult to find four harder working women in town—all good friends. Top to bottom; Mannie Salazar, Alva Gallegos, Toni Gonzales, Nellie Gallegos. *John Marshall Photo, 1994*

(below) Jose B. Gonzales (on the right) in the Mayflower Mill with David Zanoni, checking out the high grade on a separating table in its last gravity process. Toni's husband worked the Idarado, the Shenandoah, the Mayflower Mill, Peck & Glendenning, the Sunnyside. He died in August, 1992. *Zeke Zanoni Photo*

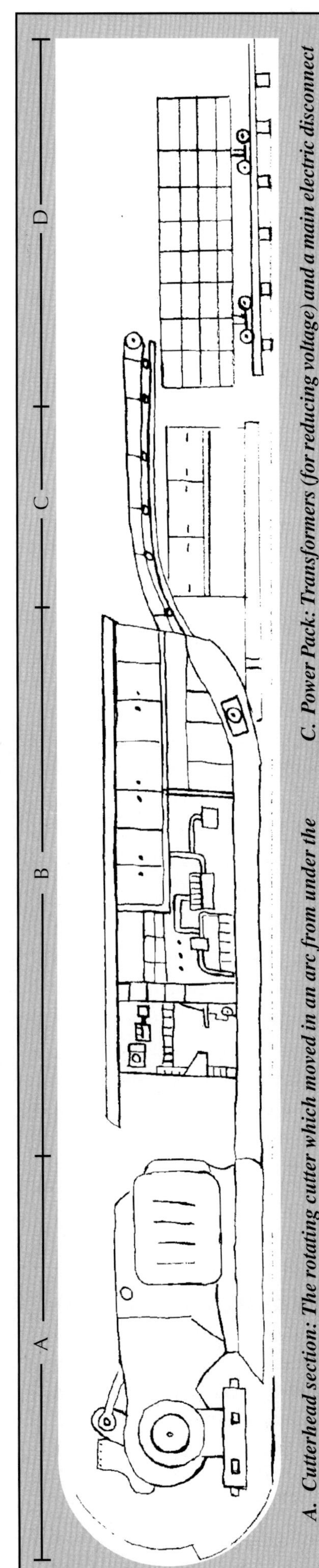

A. Cutterhead section: The rotating cutter which moved in an arc from under the machine to out in front and slightly above the machine. Chain conveyor for cuttings removal at the bottom starting just behind the rotating cutter. The round pads at the front held the cutterhead rigid and also aligned the machine left to right. The square pads at the rear held the machine in place while boring and moved about six inches horizontally between boring cycles pulling the machine forward.

B. Operator control and pumps: The operator sat in a small compartment at the front of the second section. This compartment was filled with gauges, buttons, and switches which the operator used to monitor and control the machine. Directly behind the operator compartment were mounted the hydraulic pumps, driven by electric motors and water cooled. In upper compartments towards the rear, was housed the electric circuitry. The chain conveyor ran underneath this section in a small trough which sloped upward at the rear.

C. Power Pack: Transformers (for reducing voltage) and a main electric disconnect switch were mounted in this section. The chain conveyor ran over the top, overhanging a few feet to the next section.

D. Shuttle Car: The front of the shuttle car hooked to the rear of the Power Pack section while boring operations were taking place and was filled by the cuttings carried back from the cutterhead. The car, which ran on rails, had a conveyor in the bottom that was controlled by the trammer. As cuttings were deposited in the front of the car by the main chain conveyor, the trammer would move the material to the rear of the car until it was full. The car was then pulled to the surface and discharged of its load by using outside power and the conveyor in the bottom of the car. *Jim Melcher Diagram*

Never underestimate the steps one might take to prolong and even encourage mining in the San Juans. In the end one would say it was a noble effort...

In the fall of 1976 Marvin Blackmore and myself worked for Frank Montonati, a local con-

A Tunnel Boring Machine Comes to the San Juans

tractor. Our job was to clean up an old road that was almost gone and start a portal into the mountain for a tunnel boring machine that a local mining company was bringing in from outside. The rebuilding of the road proved to be a rather difficult job as we only had a small bulldozer to work with and the dirt was frozen several feet deep. Using a hand-held sinker drill and compressor we drilled and blasted our way across the hillside finally reaching the tunnel site. It was bitter cold most of the time, and we had trouble tying the leads together on our explosives since we had to do it bare-handed.

Next a short, approximately thirty-foot, tunnel was driven into the mountain using conventional methods. This was necessary as the boring machine needed the walls of the tunnel to hold itself rigid to start boring.

The machine arrived in Silverton in February of 1977 in four big, heavy sections: the cutterhead, operator control and pumps, power pack and shuttle car. Using steel I-beams, a skid was built to haul the cutterhead. The cutterhead was so heavy that four bulldozers were necessary to move it up the mountain. Local catskinners and dozers performed this task well until a section of the newly built road gave way under the massive weight of the machine, tipping it dangerously. This caused a day of jacking and blocking before it could be pulled on up to the tunnel site.

The cutterhead was pushed into the starting tunnel, the skids removed, and the rest of the sections of the machine brought up the mountain

Jim Melcher stands here with his father Doc Melcher, who was a veterinarian for thirty-seven years. Yet, in his all too short spare time, he would come up and bug Jim for work. Here, dynamite in hand, Doc became a powder monkey. *Jim Melcher Collection*

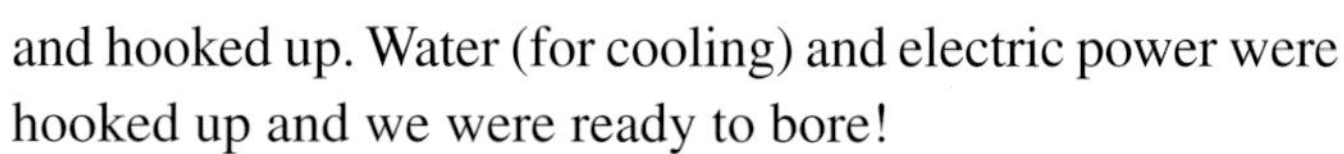

and hooked up. Water (for cooling) and electric power were hooked up and we were ready to bore!

Myself and another worker were then trained as operators of the machine by a technician from Switzerland who had been sent along with the machine.

The operation of the machine went as follows: two large pads on each side of the cutterhead section held the machine rigid while cutting by bracing against the tunnel walls. The machine also was aligned using these pads to push the cutterhead back and forth as needed to stay on course. Using this capability of movement and a laser beam mounted on the machine, one could bore thousands of feet of tunnel on grade, in a straight line. To move the machine forward, the front pads on each side were released, and the rear pads (on the cutterhead section), braced against the tunnel walls, pushed the machine ahead about six inches. The following sections being hooked to the cutterhead, were pulled along.

The five foot diameter cutterhead would be in the down position (cocked somewhat underneath the front of the machine) when the machine was being moved forward, however still rotating. After resetting the pads and realigning the machine, the cutterhead was then moved in a slow arc in front, ending slightly above the top of the machine. The cutterhead was then moved back to the down position, the machine moved forward and the process repeated. This resulted in a tunnel about five feet wide and 7° feet tall with straight sides and a curved top and bottom. This method of boring tunnels is called undercutting.

Cuttings from the boring operation were carried back to the shuttle car by means of a chain conveyor. This conveyor ran in a small compartment underneath the cutterhead and the operator/control section, then over the top of the power pack back to the shuttle car. The only access to the front of the cutterhead was through this compartment.

It would take four cats to push and pull the cutterhead to the mine. Cat skinners front to back were: Jim Melcher, Louis Girodo, Andy Archuleta, Frank Montonati. *Jim Melcher Collection*

All of this boring machine business is rather expensive and the project manager was in a hurry to get started on his project. This resulted in the machine not being set on the right upgrade when we started tunneling. This, along with the massive weight of the cutterhead and the soft surface rock encountered caused the machine to bore a tunnel slightly down hill. After about 150 feet of tunnel we were able to raise the machine to the proper grade and continue on. As we progressed into the mountain, this dip in the tunnel became filled with water about 2° feet deep. This made the laying of rail (for the shuttle car to run on) very difficult.

The machine, an Atlas Copco Mini Fullfacer #7, performed very well most of the time. When the rock was of consistent hardness with no real soft areas or extreme hard areas, one could bore considerable tunnel in a day. I was able to bore thirty-two feet in eight hours. This may have even been increased later on, as the machine was capable.

There were things that would stop it dead however, such as hard dikes like the one we hit about two hundred feet from the surface. The rock was very fine grained and hard, but luckily not very thick (about two feet). The machine would not cut this material, so it was backed up and a couple of miners were sent up to the face with jack-legs and powder. After blasting through the hard dike, the boring machine was able to continue on.

Two people were required to run the machine, the operator, and the trammer. We would bore until the shuttle car was full, and the trammer would signal me to stop boring by a light on the control panel. The shuttle car would then be hauled to the surface and its load discharged. Even though operation of the machine could be done by two people, a support team consisting of four additional miners was necessary. Track, water, and electrical had to be kept up to the machine as the tunnel progressed.

This multi-million dollar project was another attempt in the San Juans to find and develop a profitable mine at a cheaper cost and faster rate than normally done. At the last, no ore was ever milled and the funds ran out.

Before leaving Silverton in 1977, the machine bored a tunnel 1,800 feet long. The machine was quite unique, there were very few ever made and probably, like the Dodo bird, died of their own specialization. From the San Juans, the machine apparently went to Rochester, New York, to go to work boring storm water tunnels. ❋

—Jim Melcher

The operator sat in a small compartment at the front of the second section. This section is getting put on skids for the uphill trip. *Jim Melcher Photo*

The cutterhead is made ready to go into the thirty-foot tunnel already prepared for it. *Jim Melcher Photo*

This was the shuttle car that was loaded with all the broken rock. Four feet high, 3° feet wide and ten feet long, it ran on sixty pound rails. It's headed back in for the next load. *Jim Melcher Photo*

Slabbing out the crosscut in the Sultan Mine Tunnel, preparing to make room for the diamond drill cutout below. Zeke Zanoni on drill. *Zeke Zanoni Collection*

Steve Fearn outside his home in Silverton.
John Marshall Photo

When the station was ready, the diamond drill was brought in. Here, Paul Weibe (on right) and his helper Pete are diamond drilling for Boyle Brothers in the Sultan Mine, summer 1988.
Zeke Zanoni Photo

MINING, ESPECIALLY HARD ROCK MINING, IS A COMBINATION OF KNOWLEDGE AND A LITTLE LADY LUCK. It's a difficult proposition at best. Investing large amounts of money doesn't guarantee a return. How frustrating that can be. And production? It's another reason the hard rock mines are disappearing. Basic economics.

In recent years, production for one good-sized hard rock mine in the mountains of Colorado might be 1,000 tons a day, 22,000 tons a month, 264,000 tons a year. One open pit mine in Nevada can mine 240,000 tons of rock in one twenty-four hour period.

Steve Fearn speaks: "How to mine in the San Juans is pretty much a lost art. It's not taught and most people who know it are disappearing…"

There came an outfit into the San Juans in 1981 called Peck and Glendenning (P & G). The work was known as the Sultan Mountain project. Involved were the old workings of the North Star, Champion, Little Dora, Jenny Parker and the Pittsburgh Tunnel. Core drilling, sampling, reopening different parts of the mine, were the aspects of the initial development. By 1986 Steve Fearn had become General Manager. The average number of men employed over the years had been six to ten persons. That number would climb to almost one hundred men. Almost twelve million dollars had been invested. And now in 1987 operations would shut down. No ore was ever milled. The partnership was broken. Backing had ceased. The arguments would end up in court. Steve Fearn knows the gamble as well as anyone. ❋

(facing page) The old workings of the North Star Mine left from the turn of the century, were being considered for reopening. *Zeke Zanoni Photo*

The mining trade, like many of the cultures and industries within our society, has its superstitions. And of all the varieties, no single superstition has been more predominant for the miner than that of the Tommyknockers.

This deep-rooted belief in mine spirits first came to America with the Cornish miners—also known as tinners—in the 1800's. Although the concept had many social and religious connotations in Cornwall, England, as the legends of the Tommyknockers gained influence and spread through the Western mining camps of this country, they became intermingled with the beliefs and traditions of other nationalities involved in the mining labor force. This cultural melding, enhanced by the wilderness environment of the American West, gave a new twist to an already well-established superstition.

The legends of the Knockers vary greatly in detail from one source to another, but go back hundreds of years in Cornwall's history. While some of the stories depicted the Knockers as evil, in most cases they were considered a friendly lot to the miner, although sometimes mischievous. Many felt the Knockers took on the spirit of those killed in the mines. For those who believed, it was important to keep on the good side of the Knockers by speaking well of them and by leaving crumbs of food behind for their partaking.

In turn, the Tommyknockers would warn miners of dangerous situations, lead them to 'grass' (the surface) if lost or without light, and in some cases even lead them to riches beyond their wildest dreams.

The basic belief was more than just a spiritual one, for Tommyknockers were generally perceived as physical beings with great mystical powers. Descriptions have them ranging from several inches to two feet in height, with clothing anywhere from 'elf-like' to the garment of the day. Features are commonly exaggerated, and the Knockers are most often sensed as having large heads with flowing beards, long arms, short legs, and huge booted feet. These spirits have been seen and heard working in remote parts of the mines, repeating the blows of the miner's pick or sledge with great precision.
—Zeke Zanoni

Each time a man quits the sunlight to grub around underground in a mine, he enters another world.

Perhaps it isn't a part of purgatory, but I think the devil would feel quite at home down there. As the last feeble glimmer of daylight at the portal flickers out, a realm of utter blackness closes in. Damp cold fills every small tunnel and opening. The warm safety of sunshine and daylight is gone. A man will do well to accept the ways of the spirit denizens who (according to some folks) live down there.

I'm not overly superstitious. I pay little attention to ghost stories, nor do I get hung up over "taboos." However, in some ways thirteen doesn't appeal to me, though I'll bet on number thirteen on a roulette wheel any time. But never number twenty-three. I have a silly habit of knocking on wood or the side of my head when a cat crosses in front of me. It's one of those hang-overs from childhood ghost stories that made me afraid to go to bed in the dark. So now I don't have much patience with people who think there are spirits running around loose, or say there is some truth in old wives' tales.

But having to work underground from time to time, I was bound to make the acquaintance of a "Leprechaun, Elf, or Tommyknocker" the secretary for the "Spook Union" sent out to work as my helper. Learning the little fellow's name was a problem too. I used a different one each time I spoke to him until I found one he seemed to like. In my case, every time I addressed him as "Morris" (that was before the tomcat of television fame) things moved along just fine. If I neglected the formality, first one thing after another went haywire until I remembered about him, apologized, and asked his forgiveness. The little punk could be a lot of help if he was happy and willing to cooperate.

I had a real run-in with Morris one time in the old Shenandoah Mine. The company had started development on the three hundred level, and I'd been sent up there late one afternoon on an emergency repair job. The upper levels were served by a small skip which slid up and down the raise on a wooden slide. A man could ride the thing if he wished, but climbing the manway ladder was a lot safer. However, I tossed my tool sack onto the skip deck then gave old Dave the hoistman a signal to take me up to three hundred. I'd asked him to let the skip hang up there at the station, if I didn't call for it before he went off shift. That way I could lower my tools and broken parts myself, after I'd climbed down the raise to the main level hoist room.

"C'mon, Morris," I said. "Get on, let's go. I don't want to hang out here half the night."

A short time after tally I finished the job and went back to the station. Old Dave had left the skip hanging there as he said he would. On the off chance that he might have waited a little while after shift to give me a ride down, I stepped on and signaled the hoistroom "Man on, lower away."

To my surprise, the slip started down the slide quite slowly, as it should when transporting men and material. For once I was in luck. I should have known better—hoistmen never wait a minute past tally to help a repairman up out of the hole or down from the level above. I began to wonder why it happened I was getting a ride down after shift had gone off. Somebody was down there running the hoist for me, that was for sure.

By the time the skip was halfway down it had picked up too much speed. It was rattling down the slide too fast for safety. Old Dave seemed in an awful hurry to get me down from three hundred. He never slowed up. The skip hit the main level station with a good solid thump. My legs folded up—I went sprawling out onto the track in front of the hoistroom.

When I'd decided there were no broken bones between my head and heels, I scrambled to my feet, heading for the hoistroom door with just one thing in mind—knock some sense into a cantankerous old lever-man's head.

But there was nobody on the hoist that I could see, nor had there been since tally. Then I began to understand just what had happened. Morris had been running the hoist in Dave's place. Morris was mad at me because I'd forgotten to ask him to ride down on the slip. He'd whisked down the manway and set the hoist handbrake to give the skip a good hard landing when it hit the main level station. I knew he was perched up on the hoistman's chair grinning at me, so I threw a piece of rock at him. All I hit was a clock on a shelf above the hoist—smashing it to bits.

"Listen, Morris," I said, "That's enough foolishness for one day. You darn near killed me with that ride on the skip. And now I have to buy a new clock. Call it off. I'm sorry I didn't invite you to ride down with me. But you knew there was no hoistman down here and tricked me into thinking there was. Now cut out the clowning and let's be friends. I'll admit you can run that hoist as well as anyone. So why don't you do the job right once in a while? Come on, walk out to the surface with me, I'm tired and want to get down off the hill."

If I could get the little devil out to daylight, I'd hang one on him that would even the score for the day. I had made up my mind to give Morris the cussing of his life. Down the drift, daylight had just come into view at the portal, when the bottom of my tool sack split wide open. Everything fell into the drainage ditch beside the track. I had to get down on my hands and knees while I fished around in a foot and a half of muddy water to find them. The little imp could read my mind, so he'd taken one last poke at me before I could get out of his reach.

Like I said, I'm not overly superstitious, but underground where the sun never shines is another world. They have their own rules down there. Every time I got out of line Morris straightened me out in a hurry. For the most part the little pest and I made out fairly well. Several times he saved me from a serious accident with a warning any dumb bloke would understand. ❋

—Louis Wyman, from Snowflakes and Quartz

Tommyknockers living in this book are "Cousin Jack" at the top and "The Overseer" at bottom. Created by Zeke Zanoni

MOVING PEOPLE MATERIALS AND ORE AROUND IN A MINE

could be by rail, cables, air, electricity, walking, pushing, pulling, lifting, crawling—by man, machine, or mule—whatever worked.

If you were lucky you could ride the tram up. This is the Sunnyside tram at the mine in 1927. The buckets came in from the right. The rails on the left take freight back past the doghouse to various buildings around the mine. At least three of these men are on the tram crew. Phil Sartore is to the right, 1927.

Tom Knight was working the tram at the Sunnyside when the dogs on the bullwheel started wearing out. (The dogs bend over like a pair of fingers and control the cable speed. Worn out dogs cause the cable to slip and gather speed. It was the job of the men working the tram to throw rags into the bullwheel to make the cable grip and slow down. The three men in front are standing on a pile of rags.)

Tom took it upon himself to tell the 'powers that be,' "Man, you ought to replace those dogs. We've got some right here on the hill!" "No, Tom, we're making 1,000 tons a day. Not now—soon enough," was the reply he got.

It wasn't long before two bosses climbed in a bucket together for a ride up the hill. When that cable started slipping maybe nobody threw the rags... Pretty quickly those buckets seemed like they were humming along about forty miles per hour. When those two chiefs got to the top, guess what shut down and got fixed? —Richard Perino Collection

Animal power gets the ore outside in the early days. (Some of these mules never saw the daylight and grew blind.) Tramming at the Gallic Vulcan, Lake City, in the 1930s. Paul Ramsey Photo

(left) Once inside the mine, you could ride a sinking bucket up or down. Alex Johnson and Paul Ramsey hang on by their toes going to work at the Swan Mine in Rico, 1927. Paul Ramsey Collection

George Romero, hidden behind an air line and a water line, runs the tugger pulling Barney Blackmore up. Zeke Zanoni Photo

Barney Blackmore rides a coffin skip. Sure beats that ladder beside him. Zeke Zanoni Photo

(left) The skips could get pretty fancy. This was a three-tiered cage—the 44 cage in the Shenandoah-Dives Mine. Note the men in the top cage. Zeke Zanoni Collection

(below) And hey, nothing wrong with a bicycle! Buster Walker in the Idarado. These mobile units were usually reserved for staff. Zeke Zanoni Photo

Sunnyside Main Level, a fifteen-ton diesel motor, in this case pulling in a powder train to the base of the Washington Shaft. This motor also was used in pulling man trips, cars which were designed for carrying the crew to work. There could be as many as four man trip cars of various sizes that could carry as many as seventy-five to one hundred men. **Zeke Zanoni Photo**

Coming and Going in the Sunnyside

First, the miners walked into the mine, then that changed to riding in open cars called a man trip with about twenty people to a car.

There was a feeling of peacefulness, of belonging to a special group of people. Not everyone liked everyone else, in fact, there were a couple of intense dislikes, but the bosses knew and the men knew and everybody tried to work around it. If those two guys had to work together they would maybe needle each other but just like brothers they would watch out for each other. There were all kinds of ways to hurt or kill someone down there and if a guy had a brain and no conscience he'd probably get away with it.

But we all sat quietly together coming in, waiting to see what the day would bring. Different social cliques, some insistent on being as obscene as possible, gathered their members while some talked about what they saw as forces bigger than themselves that controlled their lives—like the price of gold. Some always teasing, some gently, some not so gently. Many just quietly sat with their eyes closed, often with the satisfaction of a cigarette. Of all the men on the train only a few would have their light on. Just a little light gave a mellow golden color accented with ever-changing shadows outside the train on the rib and inside across the faces. Faces with deep shadows around eyes and the angles of jaws. Smoke curling effortlessly back out of the depths of a lung and past the warmth of that body. A little rose of light that moved from what must be knee level up to mouth level where it burned more intensely for a moment and then dropped again or was suddenly snapped out of the train to die a quick death in the ditch. Sometimes a light shining outside on the wall would create some interesting geometric effects on the rock. Sometimes, I would see a profile of someone I could recognize: bushy hair, a distinct nose, something. On the whole, a pretty mellow trip in the morning. The afternoon could be a whole different story.

Such a contrast of light and dark in there. Where there was light there were two hundred watt bulbs without shades and where there were no lights, there was no light at all: pitch black, eyes cannot adjust for there is no light to see. In between was the head lamp. The rechargeable battery strapped onto our belts and a cord wrapped from battery around the shoulder to a light mounted on our hats. The light was adjusted with two

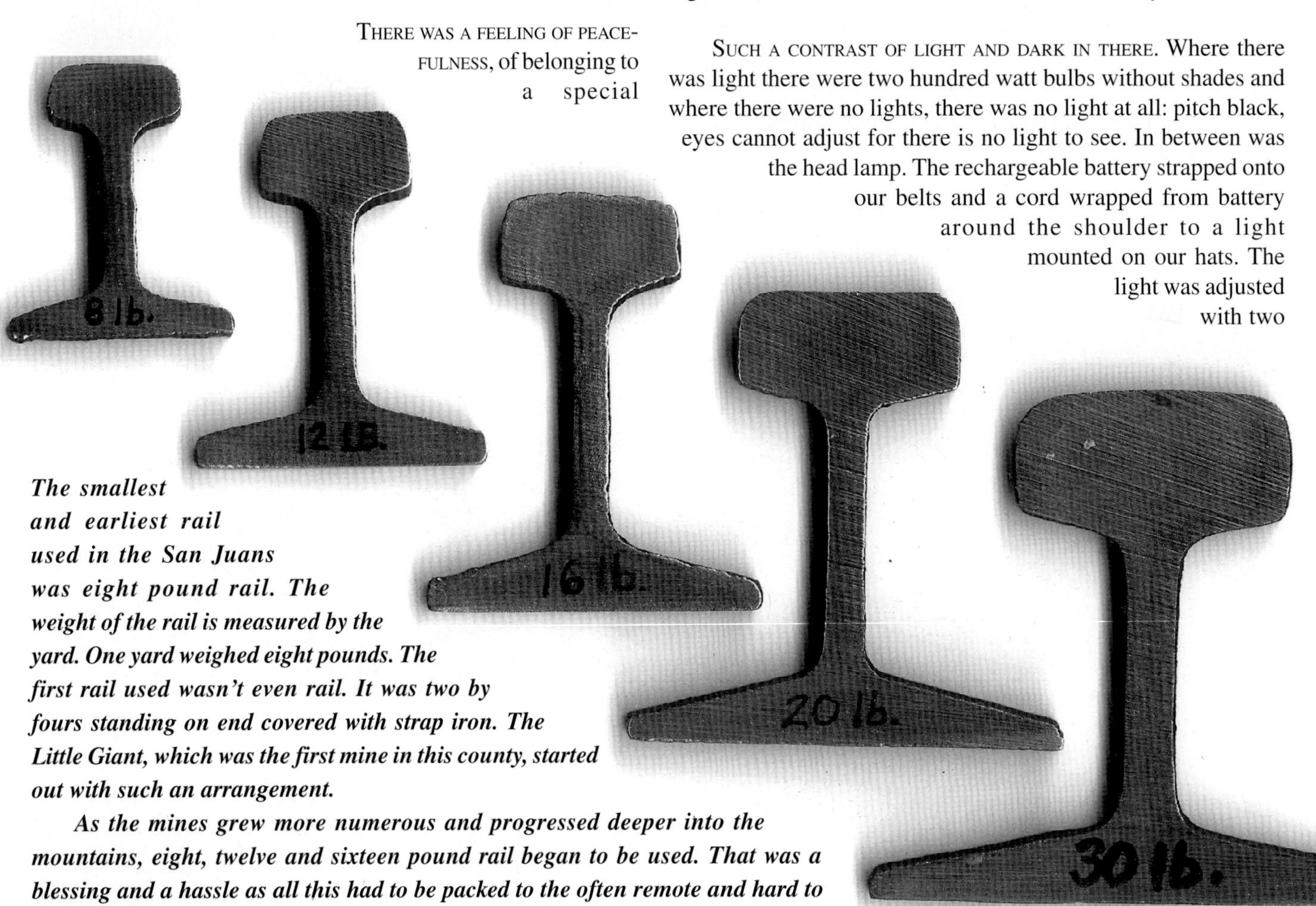

The smallest and earliest rail used in the San Juans was eight pound rail. The weight of the rail is measured by the yard. One yard weighed eight pounds. The first rail used wasn't even rail. It was two by fours standing on end covered with strap iron. The Little Giant, which was the first mine in this county, started out with such an arrangement.

As the mines grew more numerous and progressed deeper into the mountains, eight, twelve and sixteen pound rail began to be used. That was a blessing and a hassle as all this had to be packed to the often remote and hard to reach mines.

The charger station for the battery operated Mancha motors which were used for nipping (moving) supplies. *Zeke Zanoni Photo*

Randy Gallegos, tramming G-level (moving ore) in the Sunnyside. These were three-ton muck cars, loaded by ore chutes. *Zeke Zanoni Photo*

knobs, one to pick your beam, either straight ahead, flared out or off, the other to adjust the focus of the beam. Most people used the straight ahead beam. It lit less area but was more intense for careful observation. The flared beam would light up a whole area but not very well. A man without a light underground was pretty vulnerable—like a lost puppy. Sometimes we flashed them in each other's eyes. Now, even with that intense beam there wasn't all that much light so one's pupils tended to widen to take in all they could. Flash! came someone's light. Muscles contracted. The contrast could be almost painful. Sometimes folks did it to get your attention, sometimes they did it accidentally, sometimes they were signaling, sometimes they were trying to piss you off.

Light signals were as follows: side to side meant stop, up and down meant move away, and a circle meant come toward me. Our lights, more than any other piece of equipment became like an attachment of the body. Even outside the mine people would be seen shaking their head to the tune of an underground 'come to me' or 'stop.' When I would find myself in a poorly lit situation, struggling to see, I would tip my head in such a way as to get my light on the subject.

Follow that little patch of light down the drift. Stay about five feet behind it unless there was an obstruction or a stretch of particularly smooth ground. Keep it between those rails and keep moving on. And if you felt like giving somebody a little scare you just stepped off the track in a wide spot, turned off your light and waited 'til somebody came following their light down the drift. All you had to do was calmly say *"Hi"* and most people nearly jumped out of their boots. ❋

—Lois MacKenzie

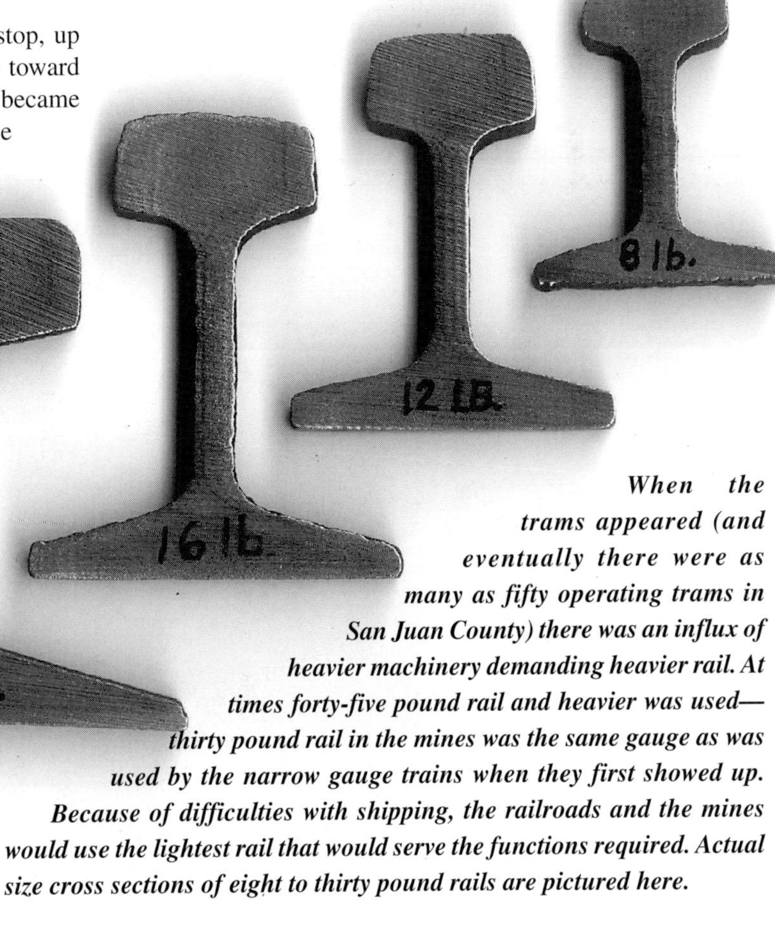

When the trams appeared (and eventually there were as many as fifty operating trams in San Juan County) there was an influx of heavier machinery demanding heavier rail. At times forty-five pound rail and heavier was used—thirty pound rail in the mines was the same gauge as was used by the narrow gauge trains when they first showed up. Because of difficulties with shipping, the railroads and the mines would use the lightest rail that would serve the functions required. Actual size cross sections of eight to thirty pound rails are pictured here.

(left to right) Archie Gurule, Donny Gurule, Roy Perino, Bill Fletcher, Terry Rhoades, Vic Gallegos and Sonny Andrean are waiting for the hoist on F-level to go to work. (Jim Gurule has his back to the camera.) Washington Incline shaft, 1983. *Zeke Zanoni Photo*

THE TRADITION OF WOMEN BEING BAD LUCK UNDERGROUND HAD ITS ROOTS IN 17TH CENTURY MINING when women and children had to work underground to keep the families fed. Few lives were considered important enough to be spared the grueling hours, deadly air and extremely dangerous ground conditions. It took a century and a half to improve conditions enough both in the outside economy and underground so that women didn't have to go under. So when a woman would show up for work, something almost instinctive would ring an alarm deep in the hearts of miners reminding them of watching their women and children dying with sweating, blackened faces, bent over from the work.

When I worked in the mine I almost had the impression the men liked having women underground. It somehow reassured them if we would work there conditions must not be all that bad. I don't remember anyone underground ever even alluding to the 'women are bad luck underground' tale.

For the most part the men didn't harass me anymore than they harassed each other. Whenever anyone new came underground the guys would push until they found the person's breaking point. If that point was too close to the surface they would drive the person out of the mine. It was a sort of survival mechanism. If pushing past the breaking point was easy, the person might not hold up in a real emergency and everyone knew their life may someday depend on the guy sitting next to them.

I watched the men drive one gal nearly nuts and decided if I was going to stay I better watch my own step. I didn't invite sexually oriented jokes or relationships, and I ran the hoist. Now, a hoist operator (hoistman in the old days) is sort of like a doctor or a cook in that you really don't want to piss them off. So when someone got rude I'd just remind them who was taking them to E-level or ask them if they wanted their feet wet in the bottom of the shaft and they usually straightened right up. Even though I threatened I never really messed around with the hoist. Too many of them had had close calls, serious accidents or lost friends in hoisting wrecks. Threats were generally effective enough.

One time only did I really scare anybody. It was the two guys who had just driven the other girl out of the mine. She had been working for them and they were on a roll. I was sent in to replace her. They got rude and after a few days began to get to me. So when they stepped on this work deck out in front of me I released the break, letting them free fall for about six feet. Then I slammed on the brake. Their saucer-shaped eyes bounced right at deck level. I asked them if they were through. They indicated they were and I never had a lick of trouble out of either of them again. They were known as the 'Fan Brothers.' One would suck and one would blow. They are both dead now. The mines took one and an unknown blood disorder took the other. I deeply respected them both. ❋ *—Lois MacKenzie*

Lois MacKenzie runs the hoist. "My boots are leaking. What a bummer. In a mine where walking through water and mud is a matter of course, a leaky boot is a bummer." *Lois MacKenzie Collection*

Mine management was always a difficult job and probably hard to talk about when you're working. Allan Bird has retired from his position as General Manager of the Sunnyside Mine under Standard Metals which he held from 1971 to 1977 and was happy to share some of his experiences. Allan is an articulate fellow, full of interesting stories. "Yes, I have three books out, and no, I don't plan any more. That's enough."

"I lived in Durango when I ran the mine up in Silverton. Mostly wise caution, I guess. My wife didn't want my kids going there. If I had to fire a guy, and now and then I did, it might have gone pretty hard on my children. I didn't want that. But you know? Yesterday I bought a house here."

•

"Oh yes, I saw all kinds of things in the mines. But the ones that always got you the hardest were the fatalities. In my time at Sunnyside there were four of them. One of the strangest ones started out down at the Antler Mine in Yucca, Arizona. On October 31, 1970, a miner fell down a shaft 450 feet to his death. Then, I left there and moved up to the Sunnyside. Exactly one year later, October 31, 1971, Max Samora—in the very same hour, almost the same minute—fell 1,000 feet to his death."

•

"Highgrading? Oh yeah, it went on. Hell, a couple of brothers bought themselves a ranch with their extra efforts. The night shift was always the worst. You just couldn't catch the guys at it. They were mighty clever. But, you know, all they could do was skim off the top couple of inches; the next seven or eight feet was the mine's. Oh, there was some pretty gold in there all right. But, hey, if the mine doesn't have enough gold for the mine and the miners, then hell, it's not a gold mine."

•

"I went to Alaska when I left Silverton, mining uranium southwest of Ketchikan on the Prince of Wales. Wherever it is, mining's a hard business. We worked five weeks on and two weeks off. I'd try to come back then to the lower forty-eight and see my wife. But that didn't work. After twenty-eight years of marriage, divorce."

The Last Mine?

The Sunnyside was going into its busiest phase and what apparently is its last phase. The seventies, eighties and nineties would see the mine now operate out of the American Tunnel built in 1963 up at the end of Cement Creek. No longer was there a town at Eureka. The mine buildings were gone from around Lake Emma. But the lake was not done with the mine—far from it. We will see that shortly. In the meantime different managers, different employees, would keep the mine in production. It was these workers and the events occurring at the mine that seemed to regulate the moods and the calendars of the people of Silverton.

Allan Bird relaxes in Silverton with his two grandsons. *John Marshall Photo, 1994*

Dale Thompson was the Sunnyside Mine Superintendent from 1977. After he got married he eased up on the coffee and twinkies. ***Allan Bird Photo***

When Dale Thompson left the mine Kay Slade took his place. At the time of this picture he was the general mine foreman. ***Allan Bird Photo***

Morning—as opposed to afternoon—was the first part of the shift before lunch, regardless of what time the rest of the world thought it was.

Allan Bird on Mine Photography

"Taking pictures underground, like anything else down there, is not easy. I liked to crawl around in the old workings and did a lot of my photography there. One time I climbed up six hundred feet. The timber had been wet since the day it was placed in 1916 and it was as good as new. I had a pack on my back and a powder bag at my side. With me was my camera, tripod, lights and what have you. It was a wet, cold climb but I finally got up there, took my pictures and made my way back down. Later on, I was back home and got that film into a little dark room I had made out of what should have been the laundry. I had a big sign on the door 'Keep Out!' Of course, halfway through developing, my daughter walked in. Oh yeah, I climbed back up that six hundred-foot shaft and took those pictures all over again."

Like so many things talked about in this book there can be a hundred different ways to work a mine and probably one could devote a book just to this particular subject. Let's make an example. We've started a mine. We've found some gold. Great! Get as much as we can and get out. Sure, that was one way to do it. But frequently there were other metals appearing with that gold. Silver, zinc, lead, and others maybe worth chasing depending on the price of metals. So let's figure our costs and take enough to cover that plus some extra. Now maybe we can make this mine last. But as we head off in different directions we run across someone else's claim. So we must work out a deal with them. They want

(left to right) Dale Rogers, Hal Slade and Billy Rhoades. Sunnyside Mine. ***Allan Bird Photo***

Terry Morris rides a Mancha Motor on F-level at the Sunnyside. ***Allan Bird Photo***

Barbara Tolleson Morris at work in the mine office, 1975. A former residence, the building was located north of town across from the tailing ponds. ***Allan Bird Photo***

"I can remember Allan taking the office help up to the mine. He dressed us up in lights, hard hats, and diggers. He brought us up to where they were working under Lake Emma. One million dollars a cut. The gold was right in front of us. Wider than your hand. A beautiful, flowing band of gold. They weren't following a vein—this was the Mother Lode! I wish I'd taken just a little piece..." —Vince Tookey, 1994

The boss's car sits outside the building where the mine offices were located. ***Allan Bird Photo***

The house of many faces. Phil Doyle told of the fifties during the uranium boom, of flying his plane into Silverton. He would land on the strip that ended close to this very house, taxi up to the door and be served a cup of tea by the woman living there. But in June of 1975, with the collapse of the #1 tailing pond, the house, then Sunnyside's office, was almost washed away. June would prove to be a very bad month for Sunnyside. Three years later all the talk would be of Lake Emma.

royalties. Okay. Eight percent if we realize this much gold. But maybe its really good ground and there's a lot of gold showing up. "Hey, I don't want you to just highgrade my mine and leave. If the heads show over a certain amount of gold I want fifteen percent royalties," says the claim owner.

To keep everybody happy and honest the ore was constantly assayed. The ore was first assayed at the mine, that way they knew what was going into the mill. And then it was assayed all through the milling process (called assaying the heads). They even assayed the tails in the last stages to make sure the recovery was as complete as it should be. So now, the guys leasing their claims and wanting royalties could tell how much gold was being extracted from their claim. The miners mining that ore knew this and therefore would mix in lower grade ores to try and keep the assay levels down to avoid higher royalties. And you thought you'd just go mine as much gold as you could!

Tailing Pond # 1 broke open on the night of June 7, 1975, washing a lot of gold into the Animas. Terry Morris was living in the office at the time and was forced to wade through neck-deep waters to safety. Photo taken from Tailing Pond #2. ***Terry Morris Photo***

In the summer of 1988, the house took a little trip—to the south end of town and became the Visitor's Center. ***Richard Bussey Photo***

DISASTER! Lake Emma Floods the Sunnyside Mine

SILVERTON COLORADO, JUNE 4, 1978

THE 2580 STOPE WAS CONSIDERED A 'HIGH GRADE' STOPE with gold assays running extremely high and with good sulfide ores as well. From the 2770 Raise pillar end to nearly the middle of the stope, on the hanging wall side, 'free gold' could be seen, sometimes in spotty patches and other times in a solid streak. It was the kind of stope every mining company dreamed of having, but it had one major drawback—it was coming up under Lake Emma. Of course this wasn't the first encounter of gold in the ore shoot on the Spur Vein. The 2590 Stope directly below, between the E & C-Levels had hit 'Free Gold' right after the pockets were 'hogged' and it was carried all the way up through that block of ground. (Refer to diagrams page 191 and 192.)

As for 2580, the company wanted to take as many six-foot cuts as they could get before shutting the stope down because of the lake. And if the vein didn't roll over into the foot wall and pinch, as had happened in the C-Level 2770 Stope next to 2580, they would probably drain the rest of the water from the lake that summer so they could mine through to the surface later in the year.

The 2580 Stope was just up nine cuts above the apex of its pocket pillars when a fault was encountered coming in on the hanging wall over 3 and 4 Pockets. At first the faulted area was small, but over a period of several days had opened up into a lens shape cavity approximately thirty feet long by four feet wide and pinching closed at the top around eight feet high. Water was showing, but on a small scale at first, of little concern. The Sunnyside was a relatively wet mine with many faults, meaning that encountering water was more of a nuisance than anything else. Yet all knew the lake was above them and it was something to think about. The crib manway had just been raised another set which was between 9 and 10 Pockets. It had been bulkheaded ready for blasting. As for the 2770 Stope end, the finger raise had been finished the week before breaking into the dog hole leading in from the raise, giving enough elevation for several more cuts. The V-cut near the middle of the stope had been drilled and shot out for the next stope cut. The stope crew had shrank several pockets so they could blast at the end of shift. The leadman, knowing that he had a ninety-foot pillar between his stope and the lake (which was considered a tremendous amount of rock), was still concerned and asked mine management to determine the status of the situation.

With a number of meetings between the stope crew, supervisors, and upper management, and the fact that the fault was still making water, it was agreed that the cut already started would be finished and the stope shut down. It was during this period of work in the stope that the wife of one of the miners had a premonition through a dream that the lake was going to flood the mine. With this, plus the tension running high anyway, the crew demanded to be taken out of the stope immediately. Realizing the urgency of the situation, the night foreman moved the crew to a temporary working place on another level until management could be notified. It was now that the scene was set for one of the most dramatic mining catastrophes in the history of the San Juan Mountains.

With the stope no longer being worked by its crew, management continued to monitor the fault on a daily basis. Everything seemed to have stabilized. The lens-shaped faulted area wasn't getting any larger and although it was still wet and dripping in some places, the small trickle of water had disappeared. The thought at that time was to leave the already broken ore in place so a stope crew would have something to work off of in the future and wait for the ice and snow to melt on and around the lake area so the trench which had been started the summer before could be deepened and the remaining water drained. This, if no problems were encountered, should eliminate the danger everyone was feeling below.

The track crew was working on Saturday the third of June. Toward the end of shift they were walking through the Spur Drift on F Level towards the Washington Shaft to go down. The track boss later reported that quite a bit of water was coming from a previously dry chute on F Level 2600 Stope. This chute, being several hundred feet lower but directly in line with 2580 Stope above, was obviously showing the first signs of what was to come. Letting the thought slip his mind, he never reported the incident. This may have been a blessing in disguise, since an investigation team might have gone to the mine Sunday with dreadful consequences.

One day later on Sunday, June 4, 1978, Lake Emma and the mud pile it was sitting on broke loose into 2580 Stope and in a matter of probably a few short hours completely inundated the Sunnyside Mine from C Level down to the American Level approximately 1,700 feet below. From there it took the only course left, by rushing through the American Tunnel and breaking out on the surface two miles away at the Gladstone Portal. Although not known at the time, the destruction of the mine was nearly complete. Knowing well the mine was in trouble, it was hoped that the Washington Shaft somehow survived, for it was the vital link between the American Tunnel and the working levels above. As was found out several months later, the hopes were in vain, for the shaft was in ruin, losing all its timber sets from G Level down, a total of over seven hundred feet. It would be two years of dangerous, backbreaking labor before the mine was back into production. Even then many areas already mined out or not needed were never cleaned out. ❋ —*Zeke Zanoni*

Sunnyside Mine, F-Level Plan View

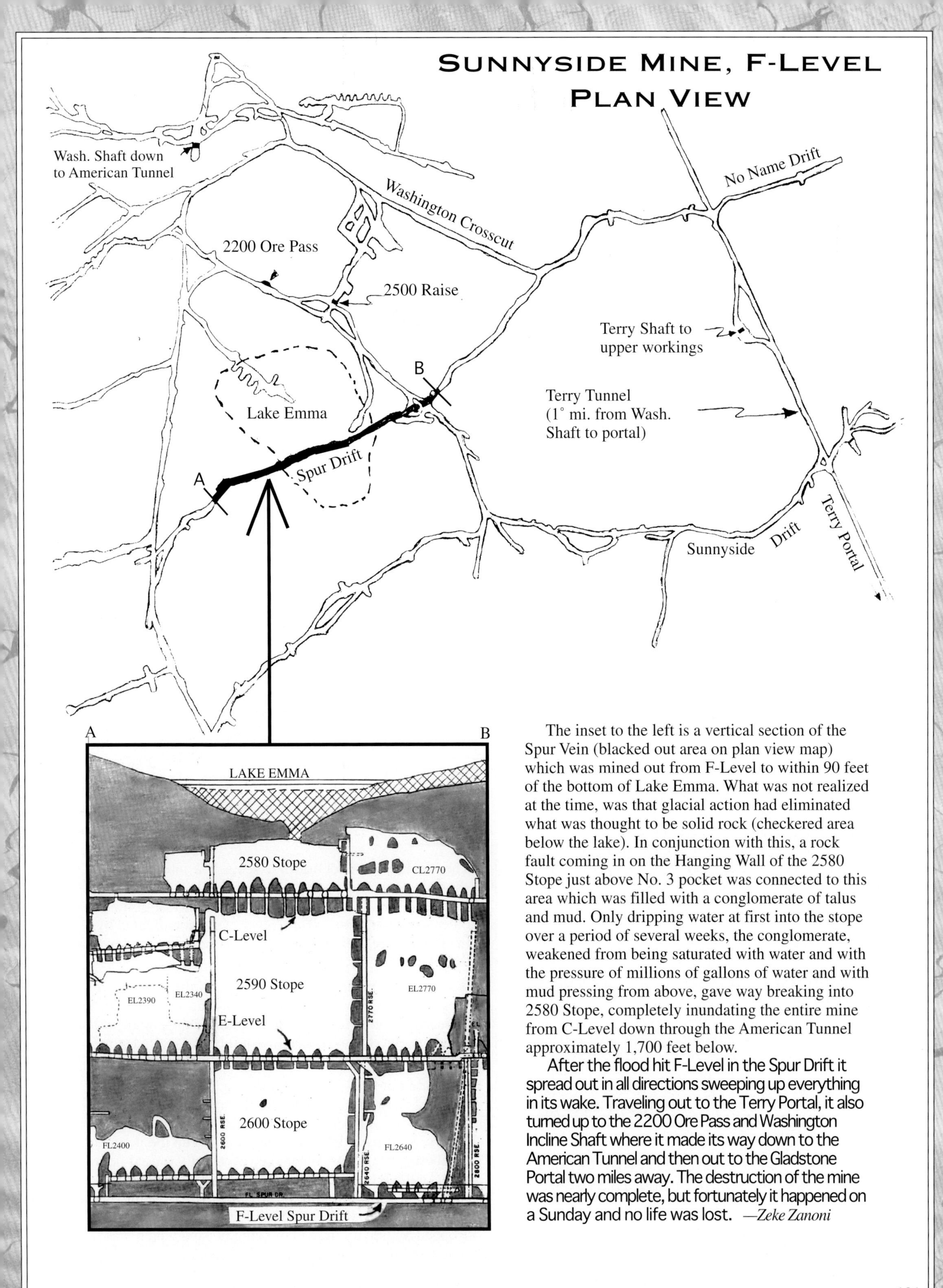

The inset to the left is a vertical section of the Spur Vein (blacked out area on plan view map) which was mined out from F-Level to within 90 feet of the bottom of Lake Emma. What was not realized at the time, was that glacial action had eliminated what was thought to be solid rock (checkered area below the lake). In conjunction with this, a rock fault coming in on the Hanging Wall of the 2580 Stope just above No. 3 pocket was connected to this area which was filled with a conglomerate of talus and mud. Only dripping water at first into the stope over a period of several weeks, the conglomerate, weakened from being saturated with water and with the pressure of millions of gallons of water and with mud pressing from above, gave way breaking into 2580 Stope, completely inundating the entire mine from C-Level down through the American Tunnel approximately 1,700 feet below.

After the flood hit F-Level in the Spur Drift it spread out in all directions sweeping up everything in its wake. Traveling out to the Terry Portal, it also turned up to the 2200 Ore Pass and Washington Incline Shaft where it made its way down to the American Tunnel and then out to the Gladstone Portal two miles away. The destruction of the mine was nearly complete, but fortunately it happened on a Sunday and no life was lost. *—Zeke Zanoni*

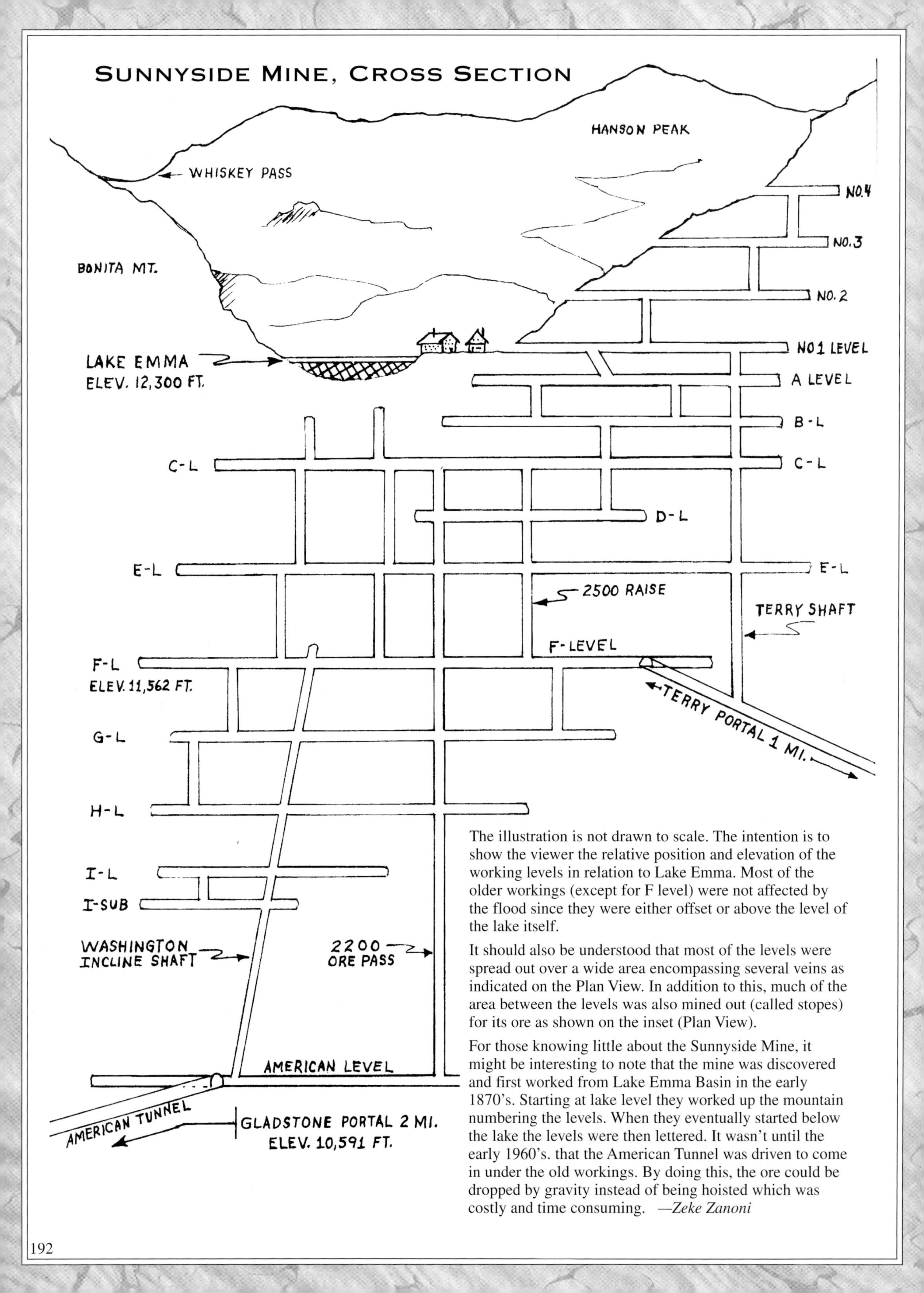

The illustration is not drawn to scale. The intention is to show the viewer the relative position and elevation of the working levels in relation to Lake Emma. Most of the older workings (except for F level) were not affected by the flood since they were either offset or above the level of the lake itself.

It should also be understood that most of the levels were spread out over a wide area encompassing several veins as indicated on the Plan View. In addition to this, much of the area between the levels was also mined out (called stopes) for its ore as shown on the inset (Plan View).

For those knowing little about the Sunnyside Mine, it might be interesting to note that the mine was discovered and first worked from Lake Emma Basin in the early 1870's. Starting at lake level they worked up the mountain numbering the levels. When they eventually started below the lake the levels were then lettered. It wasn't until the early 1960's. that the American Tunnel was driven to come in under the old workings. By doing this, the ore could be dropped by gravity instead of being hoisted which was costly and time consuming. *—Zeke Zanoni*

About halfway up Cement Creek, the second bridge here was under heavy pressure from the destructive waters, as was the road.

Zeke Zanoni Photos

Looking out of the American Tunnel Portal where the flood blasted straight out, sparing the dry room and the mine offices around to the left.

Jim Sanders is shown here in the Terry Tunnel mucking the incredibly difficult mud left behind as clean up operations painstakingly proceed.

(right) Re-timbering seven hundred lost feet of the Washington Incline Shaft. Jimmy 'Jazz' Samora on right.

Zeke Zanoni Photos

(below) On Main Level, a ten-ton ore car upended and damaged by the flood. 'Hippie John' Woodward.

At the Terry Portal. Bogged down waist deep in the mud, an old engine with no power. It took another cat to get this one out.

Cleaning out flood debris from the Washington hoist room on Main Level. Clockwise from bottom left: Frankie Gallegos, Filbert Lovato, Orlando Martinez and Max Gallegos, shifter, at bottom right. Zeke Zanoni Photos

The original break from the Lake Emma flood wasn't this big. This area's size was increased by mining after the flood, hoping to salvage some of the valuable ore. Note one of the underground levels now lying exposed.

Slowly putting things back together, Main Level mechanics shop. Left to right: Dennis Kurtz, Jim Melcher, Tom Bird (T. Bird), Bill Rhoades, Rich Perino, Trent Petersen.

From the Daily Sentinel Newspaper, Grand Junction, Colorado, June 11, 1978, by Bill Conrad.

SILVERTON—If a lot of miners are seen going to church here today, it's probably because they didn't go to work last Sunday.

Of course they will probably have a mixed reaction, since they won't be going to work for some time yet, but at least they have their lives.

What they're thankful to have missed was a wall of water that burst through the American Tunnel Sunday night, spewing mine equipment, mine timbers and debris out the portal of the mine like shot from a cannon. A witness said a wave of water five to six feet high roared from the tunnel.

The water had come from Lake Emma, eighty-five feet above the uppermost workings of the mine, when the bottom of the lake gave way and the water went down like a giant had pulled the plug of the bathtub.

Any other day of the week, even just fifteen hours later, sixty-five miners would have been working in the immediate area of the cave-in, and another eighty would have been elsewhere in the mine.

Damage to the Standard Metals Sunnyside Mine is tremendous. It is estimated that five to ten million gallons of water carried tons of tailings and lake-bottom sediments down 1,700 vertical feet through the upper four levels of the mine to the American Tunnel.

Virgil Mason, Sheriff of San Juan County, called to the scene, saw on his way up the canyon "a wave that must have been eight to ten feet high rushing down the creek. At the portal, it was like a UFO movie. Everything was black, the portal and timbers were shooting out like they were shot from a launcher."

But damage to the mine was not the end of the problem. The wall of water went on seven miles down Cement Creek to Silverton and into the Animas River.

Flood warnings were sent to residents along the river, but no serious flooding developed. What did develop was at least as serious: Pollution.

Durango, fifty miles down the Animas, was warned in time to shut off its pumps and keep the lead and zinc-polluted water out of its system. But downstream in New Mexico, Aztec and Farmington got no warning.

As a Durango resident put it Thursday, "Aztec and Farmington are really mad at Colorado today."

The La Plata County Basin Health Unit took samples from the river and got results back from Denver Wednesday. The findings were that lead and zinc exist in the river, and residents were warned not to use the water for drinking.

The lead and zinc came from tailings of old workings on the shores of Lake Emma, which was located just above Sunnyside. The tailings had been there since nineteenth century mining, and when the lake bottom fell in, the tailings washed down with the water.

At Bakers Bridge, above Durango, samples tested show zinc in the water at 12.6 parts per million and lead at 4 parts per million. Bob Balliger, acting director for Basin Health Unit, said the safe level for metals in human consumption is .05 total parts per million.

The reason for a heavy concentration of lead and zinc in the old tailings washed down by the water is that nineteenth century miners were after only gold, leaving the base metals.

Standard Metals takes gold, zinc, copper, silver and lead from the ore.

The city of Durango gets most of its water, 5.8 million gallons per day, from the Florida River, which was unaffected by the deluge, but usually takes 1.5 million gallons of water per day from the Animas.

At this time last year, the city was using an average of six million gallons of water per day.

Water from the Animas was cut off early Monday, and the city reservoir, which holds about 74 million gallons, has dropped more than two feet since then.

Lawn watering restrictions have been put into effect and citations promised to violators.

A veterinarian with the Colorado Department of Health said the water poses no threat to livestock, and agriculture specialists are checking to see if it will effect crops.

Meanwhile, Standard Metals officials are working overtime, according to The Durango Herald, to see if the $1 million dollar operation can open in the immediate future.

Miners in Silverton said the stope where the cave-in occurred was one of the richest, which is why that particular area was being mined.

The stope was eighty-five feet from the lake bottom, said D.K. Slade, mine supervisor, but Jerry Ott, Standard Metals general manager, said some other stopes were closer to the floor of the lake.

It was theorized that the lake water came down through a fault and broke into the mine.

A couple of Standard Metals employees, going on snowmobiles and on foot to the ridge above the lake, found a hole two hundred by four hundred feet around and seventy feet deep where the lake had been.

The lake is now empty, but spring runoff from the mountains around is draining into the hole, and a steady stream continues to pour from the mine.

Today, 225 miners are out of work as a result. ❋

Zeke Zanoni remembers events from eighteen years past.

Having been appointed to the position of Safety Director for Standard Metals Corporation in the summer of 1976,

the job entailed more headaches and pressure than anticipated. But, like any position, you learn the ropes and establish priorities.

As time passed, things started falling into place and the daily pressures become easier to cope with. But, each time a safety problem arose, I always felt personally responsible, even though many of the underground supervisors, I'm sure, also felt the same.

The company was making a little money, after all we had a couple of blocks of ground (ore shoots) with free gold showing, things were looking okay. Unfortunately one of the stopes having good gold values was coming up under Lake Emma. We knew it was there, and with the anticipation that the vein might go through to the surface, the company started trenching to the lake the summer before, relieving some of the pressure by draining several feet of water. The following summer, the trench would be deepened and the remaining water drained, thus eliminating any threat of a flood in the mine below.

That winter with the stope still being mined, a fault showed up on the Hanging Wall. It was small to start, with a little mud oozing out and some drips of water. But, as the days passed it become wider and longer with a small trickle of water eating away at the mud. By this time the fault had opened up into a lens shaped cavity about twenty feet long, twelve feet high and around four to five feet at its widest point.

By now the stope crew was becoming alarmed, and was demanding answers from the company. What's going on here? How much back (distance) is left from where they were to the bottom of the lake? Is this thing going to get any larger or higher? These were valid questions and we were asking them ourselves. The mine maps were being studied like they were never studied before. "We still got ninety feet of back left, we should be okay," declared someone. And we did, as far as what the existing maps showed. What we didn't know at the time, was that the fault did continue up into the bottom of the lake on one edge. Anyway, if you know anything about mining, ninety feet of solid rock is one tough chunk of ground. We had left much thinner pillars many times before with no problems. So why worry here? But, that little knot in my throat was still there, simply because you don't

This picture was taken June 5, 1978, from the American Tunnel Portal by mine manager Jim Ott.

mine under a lake everyday of the week with a fault on your hanging wall to boot!

Due to the prevailing condition, a number of meetings took place. It was finally agreed by all that the present cut which had been started in the stope would be finished and the crew moved to another working location until the lake could be drained.

It was just a few days later while on night shift that the stope crew approached the night foreman and informed him that they wanted out of the stope, and they wanted out that night. Being asked why they didn't want to finish the cut as previously agreed, one of the crew members said that his wife had had a dream the night before that Lake Emma was going to break into the stope and flood the entire mine. With the crew being very adamant in their request, the foreman had no choice but to move them to a different location until management could be advised of the predicament the following day.

The word was getting around the mine about the situation, and the talk didn't stop there. It seemed like you couldn't go downtown anymore without someone asking about the "lake thing." It was during this period of time that the General Manager called the Mine Super and myself into the office and made it very clear that we were to check the fault daily. This we did, first thing every morning. Amazingly, the entire thing seemed to stabilize, it wasn't getting any larger, and the mud seam pinched down to only a few inches. Beside this, the trickle of water had stopped to the point to where there were just a few drops here and there. To me, it was like a thousand pound weight being lifted from my back.

It was Sunday, the 4th of June. The weather had been warm and balmy for better than a week. My wife's family was down from Denver and we decided on a picnic at South Mineral. Coming into town late that afternoon, I noticed a number of local people milling around on the street corners, when we reached the Town Hall there was a large group of people standing on the front steps, most being members of the local Search & Rescue. My sixth sense told me to stop and see what was up. You haven't heard? Lake Emma broke through into the mine flooding it! Really! That is the word that came down to us from the Sheriff! Upon hearing this I immediately felt the pressure return.

My mind was racing, how could this be, the fault looked good Friday morning? Was anyone caught in the mine? I remember praying, "Please God, don't let there be anyone in the mine." It was Sunday, and the likelihood of someone underground was remote, but it was possible. My mind was like a whirlwind by this time, and I was fighting back the panic that was trying to take over. I remember saying, "Keep calm Zeke, keep calm." I was trying, but it was hard. After a couple of unsuccessful attempts on the phone to key people, I put on my hip waders and headed for the mine.

After passing through the barricade just out of town (set up to stop the general public from attempting to drive the road up to Gladstone and the mine) I kept looking down at Cement Creek which seemed to look normal. About 1° miles upstream I noticed a large wave of water riding over the top of the existing stream, and from there on instead of the murky oxidized look, it was running thick and charcoal black. It was also carrying timber and other debris as it went. The first bridge from town was okay, but as I approached the second crossing, the force of the flood was evident. So much timber and debris had been washed downstream it was plugging up the culvert forcing the thick black muddy water up and over the road, about sixteen inches deep, at one place almost washing the road away. Not being able to cross over, I stopped along side a couple of other vehicles which were already there. The Sheriff was caught on the other side coming back down from the mine. In just a few minutes, the County front end loader showed up and pushed the mud and debris from the road, at which time I climbed into another vehicle and proceeded towards Gladstone.

As we came in sight of the mine we could see where the large stacks of timber and supplies that had been stored below the portal buildings were swept away. In these areas, and where the trucks were loaded with ore, was a lake of black mud that was still moving. Surprisingly the snow sheds and surface buildings were intact. Only the snow shed directly in line with the portal and the cribbing below had been damaged, leaving everything else to either side alone. Although moving slower, mud and water was still coming from the portal and dropping down to the road level.

In order to reach the office from the approaching road where we were, we had to cross the flow of mud still coming from the mine. As I walked ahead through the mud, they followed with the truck to the other side and on up to the portal level. There we found the watchman obviously shaken and very glad to see us. Almost afraid to ask, my first question was "Did we have anyone in the mine?" With the answer coming back, "No," I felt a surge of relief almost too indescribable for words.

The first two weeks following the flood was nothing short of mass confusion. What was the company going to do? Was everyone going to be laid off? The town was choked full of outside reporters. In short, the entire atmosphere was nothing but turmoil. With the final decision being made to attempt a clean up, two years of dreadful work was about to begin to get the Sunnyside back into production. But, that's another story. ❋ *—Zeke Zanoni*

Sheriff of San Juan County, on the Flooding of the Standard Metals Mine—June 4, 1978

STATEMENT OF VIRGIL F. MASON SHERIFF

At approximately 6:50 p.m. on Sunday, June 4, 1978, after checking the lower trestle for the railroad, I arrived back in town and was met by a Mr. John Hightower, who informed me that something was radically wrong at the Standard Metals Mine up the Gladstone Road. Sheriff Mason asked him what he meant by something radically wrong. Hightower stated that there was a roar similar to an avalanche or earthquake and that water of a blackish color was coming out of the mine portal. Sheriff Mason then started up the Gladstone Road noting that the water in Cement Creek was its natural color until he reached Doug Walker's home. At this point the creek crossed the road. When Sheriff Mason passed this point there was an eight to ten-foot wave similar to water walking on the creek. Sheriff Mason stopped one hundred feet from the road to the other side. Proceeding towards the mine, Sheriff Mason noted that the water in the creek was a jet black color. Arriving at the mine gave the feeling of something out of a futuristic movie. Water was shooting out on the left side of the camelback four to five feet so that the tin on the building was standing straight out. The momentum of the water and silt was so powerful that it had picked up the timber, pipe, and other materials stored by Standard Metals Corporation in their upper and lower yards. Big timber looked like it was being shot from a launcher, thrown from the upper yard to the lower yard and into Cement Creek.

At 7:15 p.m., after reviewing the scene at the time, Sheriff Mason notified the Sheriff's Posse, the Silverton Police Department, and the San Juan Search and Rescue Team to evacuate all people on Cement Creek. In the meantime, Bill Rich had circled behind the mine on high ground where he was able to get to Brian Wogan, the mine watchman. No water or silt had found its way into the change room, the warehouse, or the mine office. Wogan decided to stay at the mine while Bill Rich returned to his vehicle where Sheriff Mason informed him to please leave the area. At this time, Chuck Williams, owner of the Mile High Antiques Store, whose cabin is above the Standard Metals Mine, came into sight and was informed by Sheriff Mason that there was no way he could cross at this time. He called on CB radio that he would return to his cabin. In the meantime, more curious people had come up the Gladstone Road and were turned back. Sheriff Mason notified Jerry Ott at this time as to what had happened. Sheriff Mason put the entire area off-limits. Brian Wogan was brought into town.

Sheriff Mason left the scene at approximately 8:30 p.m. As he started down the road, the creek bed at Walker's home as he passed, had torn up the road and filled it with debris. It was at this time that Sheriff Mason noticed the debris on the north side of the bridge had collected and was approximately fifteen to twenty feet high against the road. Water was shooting over the top of the bridge. As previously stated, the debris broke loose as Sheriff Mason passed and was piled on the road with the other side washing away, leaving only a very small passage across the bridge. Arriving at the bridge further south, it was completely under water with debris three to four feet high on the road, leav-

Looking down Cement Creek from the mine portal. June 5, 1978. *Jerry Ott Photo*

The black waters of Cement Creek, June 5, 1978. *Jerry Ott Photo*

ing Sheriff Mason stuck between these two washouts. Sheriff Mason sent for Joe Todeschi, Highway Department Superintendent, to bring up a loader. This was done, the debris removed (even though it kept filling up again) long enough to allow Sheriff Mason to cross to high ground. The creek at this time was still running very black. Sheriff Mason requested that barricades be put up. Todeschi suggested something better at the narrow part of road; that the loader be placed across the road to prevent any vehicles from passing. At this time Deputies Montonati, Hanahan, Hook and Archuleta were brought up to date and taken up the creek to the Walker home. Sheriff Mason asked Montonati if they should blast it. Jerry Stacey, Mine Superintendent for Choctaw, was sent down to get blasting powder in case it was needed. It was decided not to blast because the debris would only jam up further down the river. As long as the water was flowing, it was decided to leave it as is. All personnel left the area, Sheriff Mason putting up the barricade as he was the last to leave. This was approximately 9:30 p.m. At this time, Sheriff Mason met with Jerry Ott to determine what might have happened as to cause. It was thought at this time that Lake Emma had broken through and was emptying into the mine.

A meeting was set up at 7:30 a.m. on June 5, 1978. At the meeting, Jerry Ott, Kay Slade, Hal Slade, Pork Wilson and Ernie Kuhlman were present. It was suggested that Kay Slade and Pork Wilson try to get to the Terry while the others went up to the mine to see if they could determine the damage and what would have to be done. Returning to town, Sheriff Mason proceeded to check Eureka Creek and contact was made by radio with Hal and Pork that they had reached the Terry and that very little, if any, water was coming out. They would try to reach Lake Emma. In the meantime, Sheriff Mason returned to the Standard Metals office and met with Jerry Ott. Also with Sheriff Mason was Barry Spear of the Durango Herald. We were shown a cross section of the mine and mapping showing the exact location of the lake. At this meeting Hal and Pork came in stating they had reached the lake and nothing but a big hole was left. The hole was approximately five hundred feet long, three hundred feet wide, and going down into a cone shape about seventy-five feet. Pork and Hal described it as follows: nothing left but chunks of ice in the lake. Slade and Pork had taken one picture when they ran out of film. Sheriff Mason along with Spear returned to the Sheriff's office. After getting more details from Sheriff Mason, Spear returned to Durango.

A little after 9:00 a.m. on June 5, 1978, Sheriff Mason called Governor Lamm's office and the Division of Emergency Services, stating that we had a disaster on our hands. Full description of what had happened was given to both offices, in full detail. Sheriff Mason asked for funding if we needed it and was assured it would be available. Meantime, the State Highway Department called, volunteering any help and equipment that might be needed. County cats, a highway department cat, plus the city backhoe was used to get things started on road repairs. There was considerable damage to the highway in three different places and the road will have to be completely rebuilt in the mentioned damaged areas. Sheriff Mason received a call from Mr. Dave Lawton informing him that the state has decided to repair the road with the help of the county and that there was enough money to complete the job. This was later confirmed by Mr. John Springer of the State Highway Department.

On Wednesday, June 7, 1978, Sheriff Mason completed a trip to the mine, inspected the road, and felt it was progressing very well. The big thing of the day was that the creek was flowing in its natural brownish color. Almost all signs of black water and silt were gone, and this was also noticeable at the mine portal. At this time Sheriff Mason let it be known that on June 4, 1978, when the mishap occurred, an order was issued by him that no one was to be notified until it was determined what had caused all the problems. He also issued an order to evacuate all people on Cement Creek. ❋

NO ONE KNOWS HOW THE FLOOD LOOKED FROM THE TOP OR HOW FAST IT WENT THROUGH THE MINE. The power went out at the portal—odd for a weekend when no one was working. The

watchman, Brian Wogan, threw the breakers and sparks flew, but he still had no power. He had no phone so he wasn't able to call anyone—just figured he'd wait until he got back to town to report the problem.

Then came the sound. He thought it was a jet airplane crashing. He left his lunch and came outside. No planes. But still an incredible sound. He started for the portal.

Back in town, the first sign was Cement Creek. It started to rise and became black in color. The first people to see that, knew in their gut what had happened. And so began another era in Silverton history.

I had guests that weekend. I noticed a flurry of activity involving the Sheriff's vehicle about 4:00 p.m. but discounted it and continued enjoying the company of my friends.

The next morning those friends packed up and left. I went to the French Bakery for breakfast. Walking in the front door, I noticed about half of the people there were on day shift and should have been at work. I chided them about dumping shift. They flatly told me, "The mine caved in!"

"No way!" And they answered, "Where have you been?"

I remembered the Sheriff's lights the afternoon before. I knew this many people would never dump shift on the same day and allow themselves to be seen in public drinking coffee. Reality began to settle in.

But, no way! There was sixty-five feet or more of solid rock up there. But this group of coffee drinkers was quiet and solemn, their thoughts were spinning. I, too, sat down.

It seemed the flood of thoughts was more real than the water in the creek. Life in Silverton had changed and there was no going back. "What now? But they told us it was okay. It happened on a Sunday. What if we had been at work? What if my friends had been at work. What if I had been at work? What would we be doing now if it hadn't happened on a Sunday? How much damage was done? What next?" The cycle went round and round starting with, "No way!" All this is as we sat there quietly together.

Some people walked to the Shrine. Some in groups of two and three. Some alone. Some with their families. But all with a baptism of gratitude, the depth of which caused a physical shiver. And then the disbelief, the shaking of heads, the desire to walk away and have the whole thing never happen.

Some people went to work. It was only a couple of days before crews started working at both the American and Terry Tunnels. Indeed, life would go on. It was way too early to tell now, but eventually the mine would reopen. And produce again. But not for very long.

—Lois MacKenzie

Christ of the Mines Shrine watches over Silverton.

John Marshall Photo

Taken from The Four Corners Advisor, Silverton, Colorado, Volume 3 Number 2, Copyright 1978 by Arthur L. Francisco, June, 1978.

Due to many of you having extensive holdings in Standard Metals, most of my Advisor this month except notes will be devoted to information on the recent catastrophe.

At roughly a little after six p.m. Mountain time on Sunday, 4 June, when no one was underground, Lake Emma, at 12,300 feet crashed down through a break in its bed to the Sunnyside Mine of Standard Metals. Mud, silt and accumulation of old tailings from ninety to one hundred years ago appear to have cascaded the full extent of the mine's workings with the two mile-long American Tunnel as the exit. Emma in addition has large locations of slick rock around its perimeter and before 1908 contained mill tailings. In its course the material, a grimy black in appearance, washed out some stockpiled ore at the American Tunnel mouth, fell into Cement Creek, changing its color from red to black while substantially eroding its banks, washed out a culvert and part of a state bridge and finally joined the Animas River on its way to Durango, Aztec, New Mexico and Farmington, New Mexico.

It will be a while before the full extent of the damage is known as the American Tunnel is blocked and crews have only been able at this writing to get 1,000 feet inside the Terry Tunnel after dynamiting off its gate. Pockets of water could remain inside and precautions with this in mind are being taken.

Unfortunately much of Standard's equipment, trains, dynamite, etc. were underground at the time of the flood; hence their condition is unknown. No electrical power is available underground and many of the air compressors for ventilation are assumed to have been hit by the flood.

At a meeting on 9 June with commissioners, council and unemployment officials, mine manager Ott indicated that the work force, while repairs are underway, will be cut from 216 to approximately eighty. Thirty-six will be kept on staff, fifteen out of thirty-one in mill employees will stay on for three to four weeks to install new float cells in the mill, while out of 135 at the mine only twenty to twenty-five will stay on with chances for additional help needed in coming weeks. Approximately two days worth of ore was on hand at the mill and it shut down on 8 June after processing this ore. Company policy will, at this writing, be to spread out the clean up by working seven days per week, twenty-four hours per day with no overtime. From a $4 million dollar payroll Standard will drop by two thirds as it hopes to keep experienced people together.

Any estimate on clean up at this time is purely guess work and any reports to the contrary should be viewed in such light.

Repairs: Due to the extent of the collapse of the lake bed into an upper stope, a gouge roughly 950 to 1,000 feet long, 500 feet wide and 85 feet deep, it appears the prime step will be to divert and channel materials coming down into the hole. Spring runoff will be allowed to drain through the mine and will not be sealed off, with the possibility of the American Tunnel being such a drain.

Cash flow: There will be none from now on while damage determination and repairs are made. Standard's only source of cash flow is the Silverton mine…

In the meanwhile the common stock of Standard hit a new high of 12^1/_2$ on Thursday 8 June and closed this week at 12$^3/_8$, up $^1/_2$. (Incredible performance for a company that may have no cash flow for the next six months or more…)

What does one do? It is hoped that Standard has sufficient paid-up insurance to cover its way out of the catastrophe both as to repairs and loss of business…Be wary of anyone volunteering concrete information as to costs of damage and shutdown time, for they are not known. It will be a while and taxes in Silverton and San Juan County will go up in 1979 as the loss of over six months of Standard's production tax is felt. The best thing to close on in this necessarily long topic is to call or write your Advisor before you make any decision for new things are breaking every day. Silverton is the mining town that refuses to die! ❋

What used to be Lake Emma on the 21st of June, 1978. *Jerry Ott Photo*

And for a time in the weirdness of the markets, the stock of Standard Metals continued to rise. But reality would eventually set in.

After twenty-six years of almost continuous operation, the Sunnyside Mine under the direction of Standard Metals closed down in January of 1985, under Chapter 11 bankruptcy. This was really the beginning of the end for the famous Sunnyside Mine with its history of almost one hundred years of operations.

Although the mine and mill operated for another six years under the management of three different companies, it was losing money at every turn in the road. With the drop of base metal prices and the escalating costs of environmental obligations, it could no longer justify keeping open. In July of 1991, the production of the mine was suspended for good, laying off almost 150 employees.

From the summer of 1991 through to the present, Sunnyside Gold, with a small force of about fifteen employees, plus a number of contractors in the summer months, has continued to work its environmental commitment. Reclamation has required different kinds of work spread out over different areas. Large tailings ponds were re-contoured just outside of town. Lake Emma was smoothed over. Grass seeding and fertilization have occurred. Bulkheads were built inside the mine. There, cement plugs were poured. Discharge water from the mine has been processed previously at a lime treatment plant for years. Now, with the cement plugs in place, that water will be sealed in the mine. No one is sure of the outcome. Natural seeps may occur throughout the mountain. Hopefully, the plugs will hold. More reclamation by Sunnyside may be done on other mines whose waters enter the Animas River. Then the Sunnyside Mine will be discharged of all reclamation responsibility. The cost of this ongoing work has probably exceeded all the profits the Sunnyside ever produced in its one hundred years of mining history.

The buildings and ponds of the Sunnyside Mine stand where the town of Gladstone used to be, up by Cement Creek. Present plans call for all to be removed, erased—except for the creek. *Zeke Zanoni Photo.*

Main Level Sunnyside Mine, 1994. Left to right standing: Al Sagrillo, John Abel, Bill Goodhard, Mike Luther, Steve Legge, Dee Jaramillo, Jim Huffman, Bill McCarty, Mike Emery, Jason Perino, Jim Case, Jim Kaiser, Larry Perino, Glen Nordlander. Kneeling: Terry Rhoades, Claudia Moe, Gilbert Archuleta, Young Joe Todeschi. Scott S. Warren Photo

Young Joe Todeschi, Jim Case. ***John Marshall Photos***

Larry Perino reviews an aerial map.

Now the mine produces no ore. The miners' jobs are no longer clear cut. Each day may require a new talent.

Dee Jaramillo.

The boss, Bill Goodhard.

Mike Emery drills at home now.

The twenty-ton motor shown here has become part of the museum display in town. Joe Todeschi and Jim Case.

Al Sagrillo can't believe his pie can is empty.

Reclamation—a long, slow, arduous process—one hundred years of mining has come to this.

Scott S. Warren Photos, Gilbert Archuleta Collection.

Recontouring Tailings Pond #1, 1993.

Working on Lake Emma.

Using limestone to neutralize the ground prior to reseeding. Boyd Hadden, Mike Leghorn, Al Sagrillo.

It seems to work, 1994.

F Level, Brenneman Bulkhead—one of the first. Dee Jaramillo and Terry Rhoades, 1993.

F Level plug, Terry Tunnel. Twelve-inch stainless steel water discharge pipe. Waiting for a decision to shut off.

Store room for concrete for Terry Bulkhead, Dee Jaramillo. A 28-yard pour.

Gilbert Archuleta, 1/4-yard mixer. Two semi loads of 60 lb. bags. A continuous pour.

This was a device for pumping the concrete. American Tunnel.

Jim Huffman pouring on the Brennaman.

Larry Perino and Dee Jaramillo taking a sample for strength tests.

Larry Perino, Bill Goodhard, Jim Case and Dee Jaramillo, B Level plug.

John Wright, close to 13,000 feet on a high ridge between Hurricane Basin and Horseshoe Basin looking down Henson Creek. The Golconda Mine was there and last worked around 1920.

Andy Hanahan Collection, 1987

"IT MAKES ME SAD. I've worked the mines. I've been around them twenty years now. A guy used to tramp years ago—mine to mine, camp to camp—working where he liked and staying as long as he wanted. They were welcomed by the mine bosses, too. Because of their experience in different mines, they usually meant job safety and higher production. That's all gone now. The mines aren't there. But I still tramp—I have to. I'm headed back to Antarctica just to have a job. I go there in their summer. Explosives. Silverton winters weren't hard enough. Last year down there I remember eighty degrees below, with a wind chill of -120 degrees. I've learned a lot of difficult things in my years of mining. Been to Haiti and Central America. All over the West. And you know what? Been to Washington, D.C. twice to testify about the mining laws. I got there because friends up and down the western slope of Colorado threw in twenty-five bucks here, twenty-five there. And I was invited back a second time. Not one of the members of Congress listening had any firsthand experience of mining, yet they knew all the answers. Hmm. It makes me sad. If any place were the epitome of somewhere a guy could make a living by hard work and sweat, it was Silverton. And now that's been taken away. By legislation. By people who don't live here. Yet, I do and I love it here. It's my home. I'm building here. But be careful when you tell me what I can do. And hey, come see me, I'm in Antarctica for a while, but I'll be back..." —*John Wright*

John Wright on a warm Silverton day. ***John Marshall Photo, 1994***

The Mountains of Southern Colorado

by Dan Bender and L. Ray Liljegren, from *Friends in High Places,* copyright ©1994.

There are treasures in the mountains under Colorado skies
Many men have tried to find them and they worked hard all their lives
They found that in those mountains there is more than meets the eye
There's a peaceful existence that the world has yet to find
Now it's hard to make a going now the mining's disappeared
And the hearts of the old miners are still slowly shedding tears
And I'm stuck here in Denver but my heart is still back there
In the mountains of Southern Colorado

⚒

There are treasures in the mountains; there is still gold in the ground
But the cost of mining has gone up and the price of gold gone down
Many families have had to leave—to change their way of life
To make a living for themselves in cities with bright lights
As a boy I hiked those mountains and I fished those mountain lakes
And I saw the mountain flowers that the Good Lord chose to make
But I'm stuck here in Denver wishing I was still back home
In the mountains of Southern Colorado

⚒

There are treasures in the mountains; they are there for all to see
Many people come from miles around to view the mountains' majesty
And in the people that still live there, you can see it in their eyes
You can hear it in the way they talk you can see it in their smiles
For the mountains have affected them, have touched their varied souls
The beauty still surrounds them even through the winter's cold
And though I'm stuck here in Denver, it won't be long 'til I come home
To the mountains of Southern Colorado

⚒

I can hear the mountains calling me, their echoes fill my dreams
Not a day goes by when I don't think back on those mountains
Though I'm stuck here in Denver, it won't be long 'til I come home
To the mountains of Southern Colorado

Dan Bender with Ray Liljegren.* Friends in High Places *album cover, 1994.

Photo by Trent Peterson

A remnant Iowa tram tower, long unused, sits at the top of Arrastra Gulch and is visible for miles. *Zeke Zanoni Photo*

The Era of Mining in the San Juan Mountains is gone,
Like the passing of a friend or a loved one,
We shed a tear, and say good-bye.

Even tho' we mourn our loss,
Some of us will stay and move on with Silverton,
A town that will never be the same.
Dorothy Zanoni, 1996

Yet from dead ashes spring new fires

Preserving the history of San Juan County, Colorado. The San Juan County Historical Society Board of Directors: (front row) Zeke Zanoni, Scott Fetchenhier, Loren Lew; (back row) Allen Nossaman, Fran Schilt, Bill Jones, Bev Rich, Fritz Klinke. Loren was responsible for the construction of the archives building and is currently supervising the restoration of Town Hall. Zeke Zanoni Photo

SILVERTON IS A NATIONAL HISTORIC SITE. People do care—about each other, about the town, about their history. The museum by the Courthouse is open to visitors. A new archives building has blossomed beside the museum. Mining equipment is being added to the grounds. Perhaps, by expanding our awareness of our past, we can increase our ability to deal with the future…

The museum sits beside the Courthouse.
John Marshall Photos

We're still here. Stop by and see us. Spend some time…

and maybe remember

"There ain't no future living in the past"

But

You can't know where you're going if you don't know where you've been!

GLOSSARY

The majority of terms used in mining today came from the Cornish miner. It was his knowledge and expertise that developed most of the mines of the west. Mining in Cornwall has been traced back to before 1,000 B.C. Tin, especially, and copper were the metals they sought. Prior to steam these people had figured out how to mine down over six hundred feet using just a gad and a mallet, hand tools similar to a pick and hammer. Cornwall was the mining capitol of the world from Phonecian times. Cornish hard rock miners were revered the world over for their expertise and in the last century emigrated to North America, South Africa, and Australia among other places to make a living, frequently leaving their families back home, supporting them from a distance.

Adit or portal: An opening from the surface into a mine and/or tunnel.

Bulkhead: A structure of almost any material to hold back something. Usually heavy timber temporarily placed at the top opening of a raise or manway to stop fly-rock from entering.

Cage or mancage: A device, similar to an elevator, of many sizes and styles for raising or lowering personnel in a shaft or raise. A skip can be attached to the bottom of a cage so man and rock can be moved simultaneously.

Contract Mining: An agreement by which a company pays a mining crew for the amount of work accomplished in a given amount of time, as opposed to an hourly wage. In some cases this contract payment is for the tonnage of rock broken or per foot of advancement. Other times it was for the amount of rock moved to a designated location. In the late 1870's and early 1880's miners working in the North Star on Sultan Mountain, mostly Cornish miners, were paid under a system called ***Tutwork.*** This was a form of contract mining that had been practiced in Cornwall. Because of contract mining, Silverton, in the best of times, would be close to having the highest per capita income in the State of Colorado.

Crib: A structure composed of frames of timber laid horizontally upon one another built up like walls. One type of manway.

Crosscut: A drift or tunnel driven approximately at right angles to its destination point, and usually in barren rock to intersect a vein or other location.

Cut: In stoping, a pattern of holes drilled vertically to slab off the rock after being loaded with explosives and detonated. Usually drilled in rows from wall to wall, six feet deep. The rows are spaced about four feet apart and are drilled the entire length of the stope. After all the rows are shot out, the process is started all over again, being another cut. The cut is usually started in the middle of the stope and the rows of holes drilled back in two different directions.

Cut holes (cut): A pattern of closely drilled holes placed in the center of a round (see driving drift). They are the first series of holes fired to force out a cone-shaped hole in the rock so the other holes will have something to break to. There are many types of cuts, each with its own name such as 5-Hole burn, V-Cut, Pyramid, and so on.

Cycling a Round: This refers to the procedure and technique of mining out rock in a systematic manner. When the process is finished it begins again in the same order. Thus a miner driving drift or a raise or sinking a shaft will show up at work facing rock broken by the last shift. That rock will be mucked out, a new round is drilled and then shot as the last part of the miner's working day. Mucking, drilling, blasting, six feet at a time.

Dog hole: A small, short, horizontal excavation through a pillar, usually leading from a raise or manway into a stope or finger raise. Small enough where one must crawl through.

Dog house: Usually a small, excavated area in the mine, enclosed with boards and has a door. Furnished with table, bench and heater for mining crews to eat lunch and dry wet clothing.

Drift: A horizontal passage underground. A drift follows the vein, as distinguished from a crosscut, which intersects it.

Driving drift (drifting): To excavate horizontally. Distinguished from "sinking" a shaft down or "raising," excavating vertically upward. The technique varies, depending on its intended use. Basically, a pattern of holes are drilled horizontally (drilling a round) in the ***face*** or ***heading*** (interchangeable words referring to the end of the Drift) to a predetermined depth (usually six feet, but can vary) and as wide and high as necessary to establish size. The holes are then loaded with explosives and detonated (shot). After removing the broken rock (by a variety of methods) the process is then repeated, called ***cycling a round.***

As for sinking shaft and raising, the technique is somewhat different, but the basic principle is the same as described above in drifting, but worked vertically.

Dump, Ore: An area where commercially valuable ore is stored before shipment to mill or smelter for processing.

Dump, Waste: The waste rock dumped over the hillside outside the portal from developing a mine. This is what is seen outside most mines near the opening. No commercial value.

Engineer, Mining: Engineers who specifically measure the excavation advancement throughout the mine. Those measurements are the basis for the mining maps. These maps need to be very accurate to establish a relationship between the surface and exact locations within the underground workings.

Fault: In geology, a break in the continuity of a body of rock, attended by a movement on one side or the other of the break so that what were once parts of one continuous rock stratum or vein are now separated. The movement can be inches or thousands of feet.

Finger raise: A very small vertical raise driven short distances and just large enough to hold a ladderway. It is usually used in the stoping process to connect the stope to a larger manway.

Flotation: The modern milling process of separating the metals form the waste rock, as used in a flotation mill. After the rock is pulverized and mixed with water, the metal is literally floated to the surface with chemical reagents and retrieved. The waste rock which remains on the bottom of the float cell is carried down and out of the mill to waste piles called ***tailings ponds.***

Foot wall: The lower wall of an inclined vein. You can actually put your foot on it.

Hanging wall: Opposite the foot wall, that portion which hangs over the miner at work.

Highgrade: High Quality ore, rich in gold or silver, usually visible. ***Highgrading*** is to remove this ore from the mine, hopefully unseen, in one's pocket or otherwise—not exactly dishonorable.

Leyner (drifting drill): Technically a brand name of a rock drill (introduced shortly after the turn of the century) that was designed to be mounted on a metal column for drilling horizontal holes and in driving drift. It could be made to drill vertically as well. In later years, used as a slang word to describe any rock drill (regardless of make) which was mounted on a column.

Mancage (see cage)

Manway: Usually a vertical excavation of any size, but more often about the size of an elevator shaft, with ladders for climbing to and from working places (stopes) or to a level above. All stopes carry at least one manway. Most manways would also have a small one-man skip attached to a cable and thus to a ***hoist (tugger)*** for lifting men and material.

Moil: A sharp, pointed steel used for a variety of jobs. For example, one could hammer out a ledge in the rib in order to set a timber. The noun today means hard work or drudgery.

Nipper: A person hired to move supplies and/or material to working locations throughout the mine. Nippers usually worked in teams.

Ore chute: A structure usually made of heavy timber to catch and hold rock falling from above. Used to discharge rock into cars to be transported to another location.

Ore pass: A near vertical excavation similar to a raise or shaft, but with no timber. It is used to drop ore by gravity from upper levels to a pickup point below. In the case of 2200 Ore Pass of the Sunnyside, it starts on C Level and extends to the American Level with connections to all other levels in-between; a vertical distance of about 1,600 feet.

Ore shoot: A body of ore, usually of elongated form, extending downward within a vein. A mineralized zone of ore in a vein.

Pillar: A solid block of ground left in place to support the ***back*** (roof) or hanging wall in a stope. It can also be the rock that supports the level above a stope. Almost any block of ground between two excavations. A widely used mining term.

Raise: A vertical or near-vertical excavation driven from the bottom up and about the size of an elevator shaft, and is interchangeable with manway if used in the stoping process. At times a raise is driven from level to level, similar to a crosscut, but vertical.

Shaft: This word is probably the most misused mining term of the layman. A shaft is a vertical or near-vertical excavation, but is sunk from the top down. It can be of any size, but is usually the size of a large elevator shaft. It normally has a laddered manway and at least one large mancage which is raised and lowered with a wire rope hoist. Usually it is a main artery in a mine and can have many horizontal levels leading off of it. In one case, the Washington Incline Shaft in the Sunnyside mine is almost 1,000 feet high and intersects six levels.

Shifter: A frontline supervisor, in mine or mill.

Skip: Technically, a bucket or other device used for raising or lowering rock (ore or waste) in a shaft or raise. Usually designed to automatically dump its load at a determined point. On occasion the term is used instead of mancage. More often, skip is used to describe a one-man cage in a small raise.

Starter Steel: This short steel bar about one inch in diameter and about a foot long was used to start a hole. Depending on the length of the hole, three or more steels would follow. Starter steel, second steel, third steel, finisher, for example, for a thirty-inch hole. This technology carries over to today's modern drilling.

Stope: An excavation from which the ore has been extracted from the vein, either above or below a level. Usually mined in horizontal strips called cuts and taken up vertically. In the *shrink stope* method, only enough broken ore is extracted form the stope (shrank) while being taken up to maintain a working platform. The dimensions will vary depending on ore availability, but vertically will run from level to level and 200 to 250 feet long; the depth determined by the width of the ore. In short, a stope is a very large room with the miners working off a rock pile. When finished, the stope is *"pulled dry."*

Stoper (Buzzy): A rock drill designed to drill vertical holes. Name derived from taking up stopes. See stope above.

Stub tram: A short, aerial tram system, not unlike the larger aerial trams. Normally tied into the larger trams for carrying supplies from the main tram house to remote work areas or boarding house.

Trammer: The movement of broken rock from one location to another by various means.

Tunnel: A horizontal passage like a drift, but coming from the surface. A crosscut from the surface to intersect a vein.

References

Rocky Mountain Mining Camp - the Urban Frontier, Duane Smith; Univ. of Nebraska Press, Lincoln, 1967.

Deep Enough, Frank Crampton; Univ. of Oklahoma Press, Norman, 1982.

Colorado Gold, Stephen Voynick; Mountain Press Publ. Co., Missoula, Montana, 1992.

The Rainbow Route, Sloan & Skowronski; Sundance Publications, Denver, 1975.

Pioneering in the San Juan, Rev. George Darley; Community Presbyterian Church of Lake City, Colorado, 1976.

The Story of Hillside Cemetery, Freda Peterson; printed in Oklahoma City, Oklahoma, 1989.

An Empire of Silver, Robert L. Brown; Caxton Printers, Caldwell, Idaho, 1968.

Many More Mountains, Vol. 1 and Vol. 2, Allen Nossaman; Sundance Publications, Ltd., Denver, Colorado, 1989 & 1993.

Silverton Gold, Allan G. Bird, 1986.

Three Little Lines, Josie Moore Crum; Durango Herald, Durango, Colorado, 1960.

Bordellos of Blair Street, Allan G. Bird; Advertising, Publications and Consultants, Pierson, Michigan, 1993.

Living (and dying) In Avalanche Country, John Marshall and Jerry Roberts; Simpler Way Book Company, Silverton, Colorado, 1992.

Snowflakes and Quartz, Louis Wyman; Simpler Way Book Company, Silverton, Colorado, 1993.

Silverton Then and Now, Allan G. Bird; Access Publishing, Englewood, Colorado, 1990.

Otto Mears, "Pathfinder of the San Juan," Ruby G. Williamson; B&B Printers, Gunnison, Colorado, 1981.

Mountain Mysteries, Marvin Gregory, P. David Smith; Way Finder Press, Ouray, Colorado, 1984.

Cinders and Smoke, Doris B. Osterwald; Western Guideways, Ltd., Lakewood, Colorado, 1965.

Mountains of Silver, P. David Smith; Pruett Publishing Co., Boulder, Colorado, 1994.

Pioneers of the San Juan Country, Sarah Platt Decker Chapter, NSDAR; Family History Publishers, Bountiful, Utah, 1995.

Golden Treasures of the San Juan, Cornelius and Marshall; Swallow Press, Chicago, Illinois, 1961.

Knocking Round the Rockies, Ernest Ingersall; Univ. of Oklahoma Press, Norman, Oklahoma, 1994.

One Man's West, David Lavender; Univ. of Nebraska Press, Lincoln, Nebraska, 1943, 1956, 1977.

Father Struck it Rich, Evalyn Walsh McLean; Bear Creek Publishing Co., Ouray, Colorado, 1981.

Across the San Juan Mountains, T.A. Rickard; Bear Creek Publishing Co., Ouray, Colorado, 1980.

Of Men and Mountains, William O. Douglas; Chronicle Books, San Francisco, California, 1990.

Fun and Games of Long Ago; Chandler Press, Meynard, Massachusetts, 1988.

Queen of the Rockies, Carolyn F. Ballew; C. & D. Publishing Co., Amarillo, Texas, 1972.

Pioneering in the San Juan, George M. Darley; Community Presbyterian Church of Lake City, Colorado, 1899, 1976, 1986.

The Million Dollar Highway, Marvin Gregory, P. David Smith; Way Finder Press, Ouray, Colorado, 1986.

Blaster's Handbook; E.T. duPont de Nemours and Company, Inc. 1969.

Of Record and Reminiscence, Ruth Rathmell

Bill Bevans and little Paul Beaber, 1954. **Swanson Collection**

INDEX

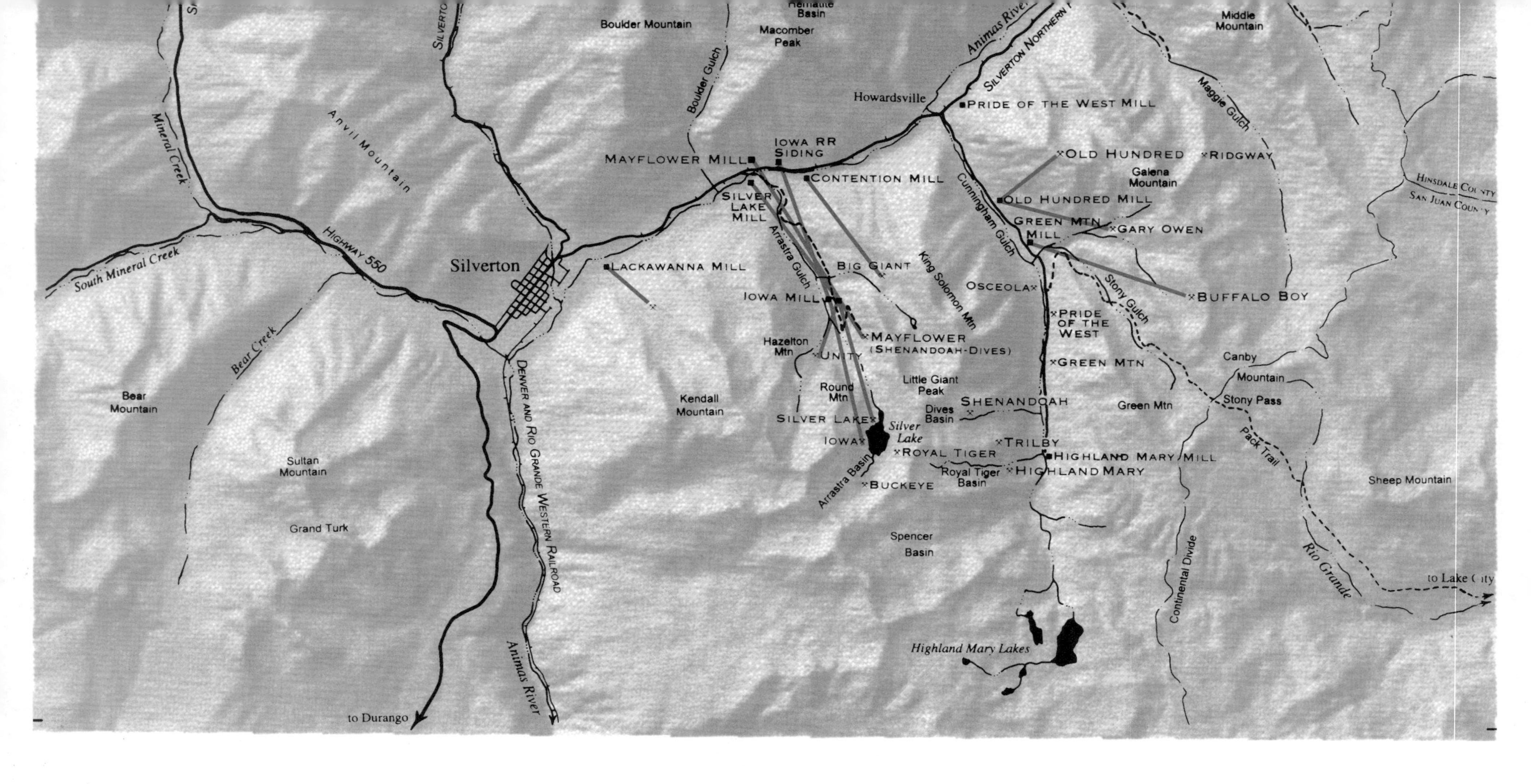

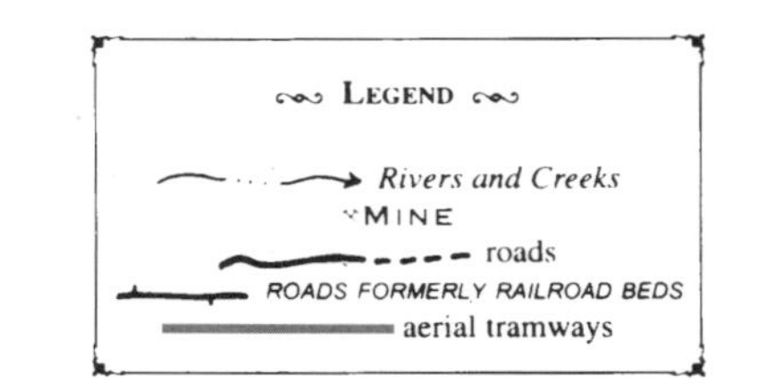

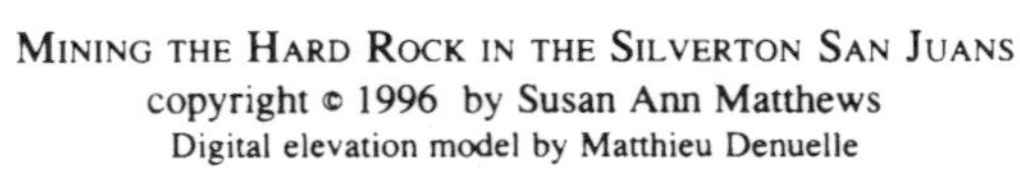

Mining the Hard Rock in the Silverton San Juans

Digital elevation model by Matthieu Denuelle

S.E.R.A.I
COMMUNICATION ARTS
473 East Third Ave · Durango, CO 81301